BrightRED Study Guide

Curriculum for Excellence

N5
PHYSICS

Paul Van der Boon

BrightRED
PUBLISHING

First published in 2013 by:
Bright Red Publishing Ltd
1 Torphichen Street
Edinburgh
EH3 8HX

Reprinted with corrections 2014, 2015 and 2016. Updated 2017.

New edition published in 2018.
Reprinted with corrections 2019.

A CIP record for this book is available from the British Library
ISBN 978-1-906736-96-5

With thanks to:
Ken Vail Graphic Design (layout and artwork) and Dr Anna Clark (copy-edit)
Cover design and series book design by Caleb Rutherford – e i d e t i c

Acknowledgements
Permission has been sought from all relevant copyright holders and Bright Red Publishing are grateful for the use of the following:

Andrey_Kuzmin/Shutterstock.com (p 6); Garsya/Shutterstock.com (p 6); graja/Shutterstock.com (p 6); ssguy/Shutterstock.com (p 6); ktsdesign/Shutterstock.com (p 6); ID1974/Shutterstock.com (p 9); mihalec/Shutterstock.com (p 9); turtix/Shutterstock.com (p 9); Ingimage (p 11); Maksim Toome/Shutterstock.com (p 16); James Steidl/Shutterstock.com (p 16); ollyy/Shutterstock.com (p 16); Paul Drabot/Shutterstock.com (p 16); anekoho/Shutterstock.com (p 21); kuppa/Shutterstock.com (p 21); Piotr Marcinski/Shutterstock.com (p 21); Alexander Sakhatovsky/Shutterstock.com (p 27); Teeraphat/Shutterstock.com (p 27); NASA (pp 30, 31, 34, 35, 36, 37, 39, 42, 43); isskh/Shutterstock.com (p 32); Galushko Sergey/Shutterstock.com (p 32); Lucky Business/Shutterstock.com (p 32); Serg64/Shutterstock.com (p 32); nikkytok/Shutterstock.com (p 32); Levent Konuk/Shutterstock.com (p 33); Igor Kovalchuk/Shutterstock.com (p 40); Igor Zh./Shutterstock.com (p 40); NASA/WMAP Science Team (p 41); Neo Edmund/Shutterstock.com (p 43); RichVintage/istockphoto (p 46); zig4photo/istockphoto (p 50); Tusumaru/Shutterstock.com (p 58); Shur23/istockphoto (p 57); Stu49/Shutterstock.com (p 59); Shutterstock.com (p 60); Carlo Toffolo/Shutterstock.com (p 61); Virunja/Shutterstock.com (p 61); GVictoria/Shutterstock.com (p 66); Lisa S./Shutterstock.com (p 67); Ingimage (p 67); Ttatty/Shutterstock.com (p 68); Blaz Kure/Shutterstock.com (p 74); photogl/Shutterstock.com (p 74); SasinT/Shutterstock.com (p 74); akud/Shutterstock.com (p 76); Konjushenko Vladimir/Shutterstock.com (p 76); Four Oaks/Shutterstock.com (p 82); R. Gino Santa Maria/Shutterstock.com (p 84); Ilya Rabkin/Shutterstock.com (p 88); Poznyakov/Shutterstock.com (p 88); Vibrant Image Studio/Shutterstock.com (p 88); Slavolijub Pantelic/Shutterstock.com (p 88); Artur Synenko/Shutterstock.com (p 89); PunyaFamily/Shutterstock.com (p 89); Christian Delbert/Shutterstock.com (p 89); leonello/Shutterstock.com (p 90); Denise Lett/Shutterstock.com (p 99); pzAxe/Shutterstock.com (p 101); lightpoet/Shutterstock.com (p 107).

Printed and bound in the UK.

CONTENTS

BRIGHTRED STUDY GUIDE: NATIONAL 5 PHYSICS

INTRODUCING NATIONAL 5 PHYSICS

The National 5 Physics course reflects Curriculum for Excellence values, purposes and principles.

Physics is the study of matter, energy and the interaction between them. This entails asking fundamental questions and trying to answer them by observing and experimenting. The answers to such questions can lead to advances in our understanding of the world around us and often result in technological improvements which enhance the lives of all.

The study of physics is of benefit, not only to those intending to pursue a career in science, but also to those intending to work in areas such as the health, energy, leisure and computing industries.

THE NATIONAL 5 PHYSICS COURSE

The purpose of the course is to develop your interest and enthusiasm for physics in a range of contexts. The skills of scientific inquiry are integrated and developed, throughout the course, by investigating the applications of physics. This enables you to become a scientifically literate citizen, able to review the science-based claims you will meet.

This course enables you to develop a deeper understanding of physics concepts and the ability to describe and interpret physical phenomena using mathematical skills. You will develop scientific methods of research in which issues in physics are explored and conclusions drawn.

The National 5 Physics course assessment has two components, a **question paper** (exam) and an **assignment**. The relationship between these two components is complementary, to ensure full coverage of the knowledge and skills of the course.

THE COURSE ASSESSMENT

Component 1 – Question Paper (80% of total mark)

The question paper is 2 hours and 30 minutes in duration in which:

- 25 marks are allocated to an 'objective test' that contains 25 multiple choice questions

- 110 marks that are allocated to the 'written paper' which includes questions requiring a mixture of short (restricted) and extended answers

The majority of marks are awarded for demonstrating and applying knowledge and understanding. The other marks are awarded for applying scientific inquiry and analytical thinking skills.

The question paper is set and marked by Scottish Qualification Authority (SQA).

Component 2 – Assignment (20% of total mark)

The purpose of the assignment is to assess the application of skills of scientific inquiry and related physics knowledge and understanding.

You will carry out research and report on a topic that allows you to apply skills and knowledge in physics at a level appropriate to National 5.

The topic should be chosen with guidance from your teacher/lecturer and must involve experimental work. The assignment allows assessment of skills which cannot be assessed through the question paper, for example the handling and processing of data gathered as a result of experimental and research skills.

contd

The assignment has two stages:

- **research**

  The research stage must involve an experiment which allows measurements to be made. You must also gather data from the internet, books or journals to compare against your experimental results.

  This stage takes place during your coursework at an appropriate time.

- **report**

  You must produce a report on your research. This will be completed under supervision.

COURSE CONTENT

The course content includes the following areas of physics:

Dynamics

In this area, the topics covered are: vectors and scalars; velocity–time graphs; acceleration; Newton's laws; energy; and projectile motion.

Space

In this area, the topics covered are: space exploration; and cosmology.

Electricity

In this area, the topics covered are: electrical charge carriers; potential difference (voltage); Ohm's law; practical electrical and electronic circuits; and electrical power.

Properties of matter

In this area, the topics covered are: specific heat capacity; specific latent heat; gas laws and the kinetic model.

Waves

In this area, the topics covered are: wave parameters and behaviours; electromagnetic spectrum; and refraction of light.

Radiation

In this area, the topic covered is nuclear radiation.

ONLINE

This book is supported by the BrightRED Digital Zone - head to www.brightredbooks.net/N5Physics for videos, quizzes, games and more!

 ## THINGS TO DO AND THINK ABOUT

This book will guide you through the content and skills you need to succeed at National 5 Physics. So, let's get started!

DYNAMICS

VECTORS AND SCALARS 1

The key concepts to learn in this topic are:
- the definition of vector and scalar quantities
- the identification of force, speed, velocity, distance, displacement, acceleration, mass, time and energy as vector or scalar quantities

- the calculation of the resultant of two vector quantities in one dimension or at right angles
- the use of appropriate relationships to solve problems involving velocity, speed, displacement, distance and time:
$$s = vt, \ s = \bar{v}t, \ d = \bar{v}t$$

Temperature is defined by value and unit

Scalar quantities:

distance

mass

Vector quantities:

velocity

weight

SCALARS AND VECTORS

Physical quantities are studied in physics. Physical quantities can be classified as one of two types:

- scalar quantities
- vector quantities

It is important to be able to identify scalar and vector quantities.

Scalar quantity

A scalar quantity is defined by its *magnitude* alone. This means that it only has size. Scalar quantities can be added or subtracted using basic arithmetic. For example, the temperature of an object is a quantity which has a value and a unit (°C). This information is all that is required to define this quantity.

Vector quantity

A vector quantity is defined by both its *magnitude* and its *direction*. This means that for a vector both size and direction must be stated. For example, the direction could be up or down, left or right, or the direction of a compass bearing such as north east (NE) or south (S).

Vectors *cannot* usually be simply added or subtracted; the direction is important and has to be taken into account. To add vectors together, we have to use scale diagrams or mathematical formulae, such as trigonometry and Pythagoras' theorem.

Some scalar and vector quantities used in this course are shown in the tables below.

Scalars	Symbol	Unit
distance	d	m
speed	v	$m\,s^{-1}$
time	t	s
mass	m	kg
energy	E	J
power	P	W
temperature	T	°C
work	E_w	J

Vectors	Symbol	Unit
displacement	s	m
velocity	v	$m\,s^{-1}$
acceleration	a	$m\,s^{-2}$
force	F	N
weight	W	N

DON'T FORGET ✚

You should know which quantities are scalars and which are vectors.

ONLINE ➡

For a whole lot more on scalars and vectors, have a look at 'Describing Motion with Words' at www.brightredbooks.net/N5Physics

DISTANCE AND DISPLACEMENT

Distance

Distance is a scalar quantity. The symbol for distance is d. Distance is defined by a number and its unit, the metre, m. For example, 'the length of the school laboratory is 8 m'.

Displacement

Displacement is a vector quantity. The symbol for displacement is s. Displacement is the direct distance travelled in a stated direction from the starting point. For example, 'an explorer walks 20 km due north from the base camp'.

contd

EXAMPLE 1

A student travels 10 m east along a straight corridor and is sent back 7 m for running.

(a) What is the distance travelled by the student?
Total distance travelled, $d = 10 + 7 = 17$ m

(b) What is the displacement of the student from the starting point?
Displacement is $10 + (-7) = 3$ m east

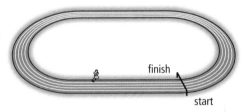

EXAMPLE 2

A runner goes round a race track three times. The track has a length of 220 m.

(a) Calculate the distance travelled by the runner.
Distance travelled, $d = 3 \times 220 = 660$ m

(b) Determine the final displacement of the runner from the starting point.
Displacement, $s = 0$ m (since the runner finished 0 m from the start line)

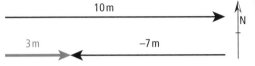

Speed

Speed and average speed are scalar quantities. The symbol for speed is v and the symbol for average speed is $\bar{v}$. Speed is the distance travelled by an object per unit of time. The unit for speed is metres per second, m s^{-1}. Speed does not have a direction.

The relationships used to calculate speed and average speed are: $d = vt$ and $d = \bar{v}t$.
Where: v = speed $\bar{v}$ = average speed d = distance travelled
    t = time taken to travel this distance.

In example 1(a) above, if the student took 5 seconds to travel the total distance of 17 metres, then, average speed $\bar{v} = \dfrac{d}{t} = \dfrac{17}{5} = 3.4 \text{ m s}^{-1}$.

In example 2(a) above, if the runner took 150 seconds to travel the total distance of 660 metres, then, average speed $\bar{v} = \dfrac{d}{t} = \dfrac{660}{150} = 4.4 \text{ m s}^{-1}$.

Velocity

Velocity and average velocity are vector quantities. The symbol for velocity is v and the symbol for average velocity is $\bar{v}$. Velocity is the displacement of an object per unit of time including the direction of the object with respect to its starting position.

The relationships used to calculate velocity and average velocity are: $s = vt$ and $s = \bar{v}t$

Where: v = velocity $\bar{v}$ = average velocity
    s = displacement from starting position
    t = time taken to travel this displacement.

In example 1(b) above, if the student took 5 seconds to travel a displacement of 3 metres east, then, average velocity $\bar{v} = \dfrac{s}{t} = \dfrac{3}{5} = 0.6 \text{ m s}^{-1}$ east (note direction **is** required for non-zero velocity).

In example 2(b) above, if the runner took 150 seconds to travel a displacement of 0 metres, then, average velocity $\bar{v} = \dfrac{s}{t} = \dfrac{0}{150} = 0 \text{ m s}^{-1}$ (no direction because runner returns to the starting position)

The unit for velocity is metres per second, m s^{-1}, and the displacement is described using a three-figure bearing or by giving direction points of the compass. For example, the velocity of the bus was 18 m s^{-1} at a bearing of 315 (or N 45° W).

DON'T FORGET

Displacement is different from distance. Displacement is the distance from start to finish in a direction.

DON'T FORGET

You must include the direction when giving a vector answer.

ONLINE TEST

How well have you learned about scalars and vectors? Take the 'Velocity and displacement' test at www.brightredbooks.net/N5Physics

ONLINE

For a further example on velocity and displacement, go to www.brightredbooks.net/N5Physics

THINGS TO DO AND THINK ABOUT

Find the displacement of your house from your school or college. What information do you need? Remember, the displacement requires a direction as well as the actual distance. Use the time it takes for you to reach school to calculate average speed and average velocity.

VELOCITY AND SCALARS 2

The key concepts to learn in this topic are:
- the calculation of the resultant of two vector quantities in one dimension or at right angles
- the determination of displacement and/or distance using scale diagram or calculation

- use of appropriate relationships to solve problems involving velocity, speed, displacement, distance and time:
$s = vt$, $s = \bar{v}t$, $d = \bar{v}t$.

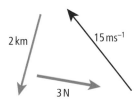

2 km

15 ms^{-1}

3 N

Vectors (not to scale)

COMBINING VECTORS

Vectors can be represented by a straight line with an arrowhead. The size of the vector is represented by the length of the line. The direction of the vector is represented by the direction of the arrow.

Vectors can be added or combined to give a single vector which is known as the *resultant* vector. When more than one single vector acts in the same straight line (or in one dimension) they can be added algebraically.

EXAMPLE 1

A space rocket has a weight of 15 000 N. At lift-off, the engine force is 25 000 N. Calculate the resultant force.

The forces are vectors:
- the weight (15 000 N) acts downwards
- the engine force (25 000 N) acts upwards.

Resultant force = 25 000 – 15 000 = 10 000 N upwards

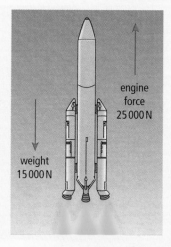

engine force 25 000 N

weight 15 000 N

Stating the direction of the resultant

The direction of the resultant should be stated in the same terms as the original information. For example, in Example 1, the force vectors clearly act upwards or downwards. When compass directions are given, the direction can be stated as a three-figure bearing or as a compass bearing.

N

This is how the north direction is usually indicated on diagrams

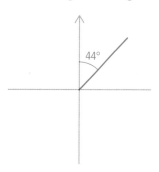

Three-figure bearing: 044 (include first zero to make three digits, no degrees symbol needed)
Compass bearing: 44° east of north

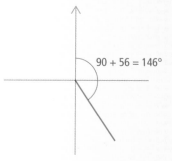

Three-figure bearing: 146
Compass bearing: 56° south of east

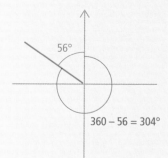

Three-figure bearing: 304
Compass bearing: 56° west of north

Direction expressed as equivalent three-figure bearings and compass bearings

When there is an angle between the single vectors, then a scale drawing or mathematical calculation (using Pythagoras' theorem and trigonometry) has to be used to determine the resultant.

contd

For scale drawings, these steps set out the process

- Choose and write down a suitable scale to represent the vectors in the diagram.
- Draw the direction indicating north.
- Using a ruler, draw the vectors according to the chosen scale.
- Add vectors 'head to tail'.
- Draw the resultant from start to finish.
- Measure the length of the resultant.
- Use the scale to convert back to the vector quantity.
- Measure the angle of the resultant.
- State this as a three-figure bearing.

You will need this equipment to draw scale diagrams

VIDEO LINK

Check out the 'Vectors' link at www.brightredbooks.net/N5Physics

EXAMPLE 2

An athlete runs 400 m due west then 300 m due north. By scale diagram, or otherwise, determine the resultant displacement of the athlete.

Solving example 2 by scale diagram

Resultant length

= 10 cm

→ 10 × 50 = 500 m

Resultant angle = 37°

Note the 37° is in 4th quadrant of a compass.

So three figure bearing is 270 + 37 = 307

Resultant displacement is 500 m at a bearing of 307 (or 53° west of north).

Solving example 2 by calculation using Pythagoras' theorem and trigonometry:

$x^2 = y^2 + z^2$

$x^2 = 400^2 + 300^2 = 250\,000$

$x = 500$ m

$\tan\theta = \dfrac{300}{400}$ $\theta = 37°$

Resultant displacement is 500 m at a bearing of 307, (or 53° west of north).

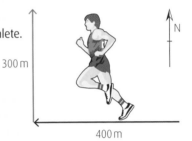

Solving example 2 by scale diagram

Average speed and velocity

In example 2, if the athlete took 200 seconds to complete the exercise, then total distance travelled, d = 400 + 300 = 700 m,

and average speed, $\overline{v} = \dfrac{d}{t} = \dfrac{700}{200} = 3\cdot5$ ms^{-1}

the average velocity, $\overline{v} = \dfrac{s}{t} = \dfrac{500}{200} = 2\cdot5$ ms^{-1} at a bearing of 307.

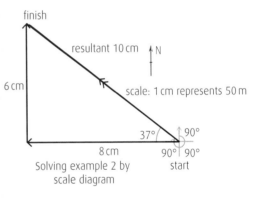

Solving example 2 by calculation

 THINGS TO DO AND THINK ABOUT

1. Investigate the sport of orienteering. Find out how the contestants navigate their way around the course.

2. What is the importance of vectors in this sport?

VELOCITY AND SCALARS 3

The key concepts to learn in this topic are:
- calculation of the resultant of two vector quantities in one dimension or at right angles
- determination of velocity and/or speed using scale diagram or calculation

- use of appropriate relationships to solve problems involving velocity, speed, displacement, distance and time
$s = vt$, $s = \bar{v}t$, $d = \bar{v}t$
- description of experiments to measure average and instantaneous speed.

EXAMPLE 1

During an orienteering race, a contestant runs from start to finish in a total time of 2·4 hours. From the starting position, the contestant runs 2 km north, then 4 km east then 2 km south then 2 km west to finish.

(a) **Calculate the contestant's distance travelled.**
The actual **distance** travelled is the sum of each part of the race
$d = 2 + 4 + 2 + 2 = 10$ km.

(b) **Calculate the contestant's displacement.**
The **direct distance** of the finish from the start is 2 km.
The **direction** of the finish from the start is **east**.
So displacement s = 2 km east.

(c) **Calculate the contestant's average speed**
Average speed is calculated using
$\bar{v} = \dfrac{d}{t} = \dfrac{10}{2\cdot4} = 4\cdot2$ kmhr^{-1}

(d) **Calculate the contestant's average velocity**
Average velocity is calculated using
$\bar{v} = \dfrac{s}{t} = \dfrac{2}{2\cdot4} = 0\cdot8$ kmhr^{-1} east

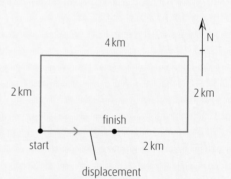

EXAMPLE 2

A helicopter travels due east at a velocity of 48 ms^{-1} and encounters a crosswind of 12 ms^{-1} due south.

Determine the resultant velocity of the aircraft.

Solving by scale drawing method

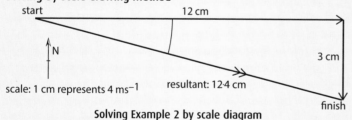

scale: 1 cm represents 4 ms^{-1}

Solving Example 2 by scale diagram

Resultant length = 12·4 cm
$\rightarrow$ 12·4 × 4 = 49·6 ms^{-1}

Resultant angle = 14°,
so three-figure bearing is 90 + 14 = 104

Resultant velocity is 49·6 ms^{-1}
at a bearing of 104, (or 14° south of east).

How to solve problems using Pythagoras' theorem and trigonometry

Solving using Pythagoras' theorem and trigonometry:

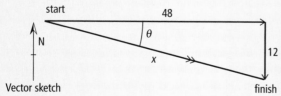

Vector sketch

$x^2 = y^2 + z^2$

$x^2 = 48^2 + 12^2 = 2448$

$x = 49\cdot5$ ms^{-1}

$\tan\theta = \dfrac{12}{48}$ $\theta = 14°$

The resultant velocity is 49·5 ms^{-1} at a bearing of 104, (or 14° south of east).

contd

Measuring average speed

Average speed refers to the speed of an object over the total distance travelled during a journey.

The relationship used to calculate average speed is $\bar{v} = \dfrac{d}{t}$

Three things are important when describing how to measure average speed:

1. measure the distance, d, travelled by the moving object

2. measure the total time taken, t, to travel this distance

3. use $\bar{v} = \dfrac{d}{t}$ to calculate the average speed.

EXAMPLE 3

Describe an experiment to measure the average speed of a car

1. measure the distance between two markers (e.g. lampposts)

2. use a timer to measure the time to travel this distance

3. calculate the average speed using $\bar{v} = \dfrac{d}{t}$

stopwatch   measuring tape

Measuring instantaneous speed

Instantaneous speed refers to the average speed of a moving object over a **small** time interval (usually less than one second).

Three things are important when describing how to measure instantaneous speed:

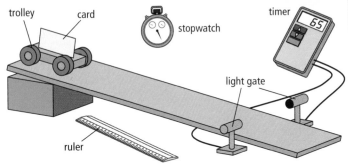

trolley   card   stopwatch   timer   light gate   ruler

1. measure the distance travelled by the moving object over a **short** time interval

2. measure the time taken to travel this distance

3. use $\bar{v} = \dfrac{d}{t}$ to calculate the instantaneous speed.

An electronic timer is usually used to measure the short time interval.

EXAMPLE 4

Describe an experiment to measure the instantaneous speed of a trolley on a ramp as it passes through a light gate

1. measure the width, d, of the card on the trolley

2. record the time taken, t, for the trolley to pass through the light gate

3. use $\bar{v} = \dfrac{d}{t}$ to calculate the instantaneous speed

EXAMPLE 5

Students ride on a roller coaster carriage which passess through a water hazard feature.

Describe a method to determine the instantaneous speed of the carriage as it reaches the water hazard.

Set up a light gate and timer at the beginning of the water hazard. Measure the length, d, of the carriage.

Record the time, t, on the timer when the carriage passess through the light gate.

Use $\bar{v} = \dfrac{d}{t}$ to calculate the average speed of the carriage.

VELOCITY-TIME GRAPHS 1

The key concepts to learn in this topic are:

- drawing or sketching of velocity–time or speed–time graphs from data
- interpretation of a velocity–time graph to describe the motion of an object
- determination of displacement from a velocity–time graph:

$$s = \text{area under } v\text{–}t \text{ graph.}$$

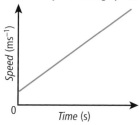

Constantly increasing speed

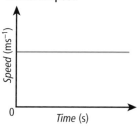

Constant speed

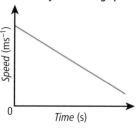

Constantly decreasing speed

Different types of motion represented by speed–time graphs

VIDEO LINK

Check out the 'Speed-time graphs' clip at www.brightredbooks.net/N5Physics

DON'T FORGET

The shape of a speed-time graph describes the object's motion.

SPEED-TIME GRAPHS

Speed–time graphs are often used to display motion, which often is not in a straight line.

Uniform **acceleration** means that the object's speed increases by the same amount each second or decreases by the same amount each second (sometimes called **deceleration**).

EXAMPLE 1

The diagram shows a speed–time graph for a Formula 1 car completing part of one lap of a racetrack.

A Formula 1 racetrack has right and left turns and long, straight stretches of track. So, the driver has to change

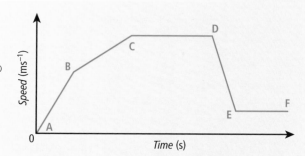

the direction of the car many times during one lap. The car speeds up, slows down, and sometimes travels at constant speed.

Use the letters on the graph to identify the section(s) which illustrate:

(a) positive acceleration
A–B and B–C

(b) constant speed
C–D and E–F

(c) negative acceleration (deceleration).
D–E

EXAMPLE 2

Which of the following speed–time graphs represents the motion of a vehicle that accelerates from rest, then moves at constant speed, then accelerates again, then decelerates to rest?

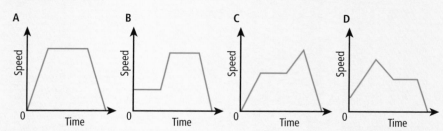

C is the correct answer. A shows only one acceleration. B starts at a constant speed, not from rest. D has two decelerations.

When drawing a speed–time graph, always label the origin of each axis with zero and choose scales carefully to produce a graph which is a reasonable size. Remember to label each axis with its name and the unit.

contd

DON'T FORGET

In speed-time graphs, the direction is not considered, even if the train is reversing.

EXAMPLE 3

A metro train is at rest at station A. The train accelerates uniformly from rest to a speed of $35\,ms^{-1}$ in 25 s. It continues at this speed for 45 s, passing through station B without stopping. The driver then applies the brakes for 20 s, bringing the train to a stop at the end of the line, station C, where it waits for 20 s. The train then returns in the opposite direction. It accelerates uniformly from rest to reach a speed of $30\,ms^{-1}$ in 20 s. The driver then applies the brakes for 15 s to stop at station B.

Draw the speed–time graph of the train.

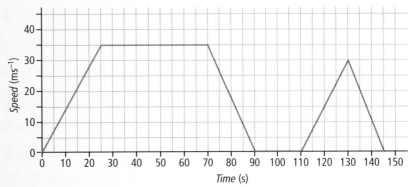

EXAMPLE 4

The speed of a trolley is recorded at regular intervals as it moves down a runway. The first part of the runway is steeper than the second part.

The results are recorded in the table.

Speed (ms^{-1})	0·3	0·6	0·9	1·2	1·5	1·8	2·1	2·2	2·3	2·4
Time (s)	0·1	0·2	0·3	0·4	0·5	0·6	0·7	0·8	0·9	1·0

(a) Draw a speed–time graph of the results.
See graph opposite.

(b) Describe the shape of the graph.
The graph has two different accelerations, greater at the start than at the end. This is seen from the different gradients of the graph – the first section has a steeper gradient, which indicates a greater acceleration.

(c) Explain the shape of the graph.
The slope of the second part of the runway is less steep than the first part. The trolley will have a smaller acceleration on this part of the slope.

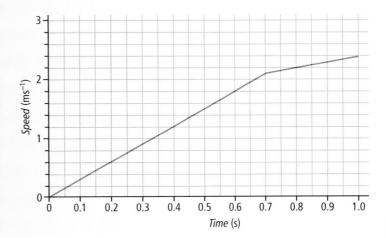

THINGS TO DO AND THINK ABOUT

During maintenance work on motorways, the speed of vehicles is usually restricted. Average speed cameras can be used to control traffic speed. Investigate how this process works.

1. How is the car recognised, timed and the average speed calculated? Is it a successful method of speed control?

2. Can cars 'dodge' the process by travelling very fast on some stretches and going very slow on other parts of the restricted stretch of road?

ONLINE TEST

Take the 'Graphs and calculations' test at www.brightredbooks.net/N5Physics

VELOCITY-TIME GRAPHS 2

VELOCITY-TIME GRAPHS

Velocity–time graphs can be used to describe the motion of objects travelling *in a straight line.*

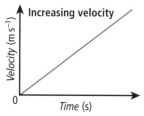

positive acceleration
The gradient of the slope of the graph is positive – this means that the acceleration is positive. The vehicle is moving forwards.

These graphs represent the motion of a vehicle travelling on a straight road

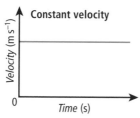

no acceleration (0 ms⁻²)
Constant positive velocity, always above the time axis. The vehicle is moving forwards.

The following velocity–time graph represents the motion of a car which travels in one direction, then returns in the opposite direction.

Note that motion in the forward direction is represented by the positive section of the velocity–time graph and motion in the reverse direction is represented by the negative section.

deceleration
The gradient of the slope of the graph is negative – this means that the acceleration is negative.
There is a positive initial velocity above the time axis and a final negative velocity below the time axis.
The section above the time axis shows that the object was moving forwards.
The section below the time axis shows the object slowed down to stop ($0\,\text{ms}^{-1}$) then increased velocity in the opposite direction.

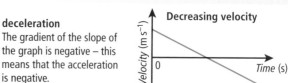

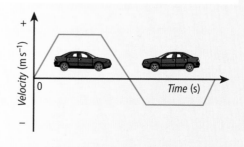

EXAMPLE 1

Study this velocity–time graph for an object.

(a) During which times is the velocity of the object positive?
During 0–3 seconds the velocity is positive (above the time axis).

(b) During which times is the velocity of the object negative?
During 3–8 seconds the velocity is negative (below the time axis).

(c) When is the velocity $0\,\text{ms}^{-1}$?
At 0, 3 and 8 seconds.

(d) During which times is the acceleration of the object zero?
1–2, 4–5, 6–7 seconds for zero acceleration (horizontal line means constant velocity).

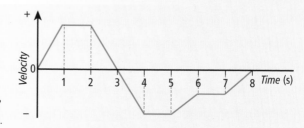

(e) During which times is the acceleration of the object positive?
During 0–1, 5–6 and 7–8 seconds there is positive acceleration (the gradient of slope is positive).

(f) During which times is the acceleration of the object negative?
During 2–4 seconds there is negative acceleration (the gradient of the slope is negative).

SPEED-TIME GRAPHS AND DISTANCE TRAVELLED

The area under a speed–time graph is equal to the distance travelled by a moving object.

EXAMPLE 2

A car's journey is represented by the following speed-time graph. Calculate the total distance travelled by the car.

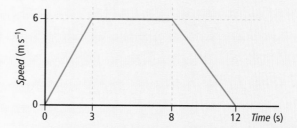

contd

The speed–time graph is split up into triangular and rectangular shapes to allow the areas of each shape to be calculated.

The distance travelled by the car during the first 3 s of the journey is the area of the first triangle:

area = ½ base × height = 0·5 × 3 × 6 = 9 m

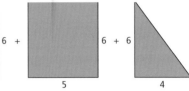

The distance travelled during the whole journey lasting 12 s would be the total area under the graph. The area is broken down into two triangles and a rectangle.

Total area = (0·5 × 3 × 6) + (5 × 6) + (0·5 × 4 × 6) = 9 + 30 + 12 = 51 m

EXAMPLE 3

The following is a speed–time graph for a bus travelling between two stops.

(a) Calculate the distance between the bus stops.

The graph must be divided into shapes for calculating the total area: three triangles and one rectangle.

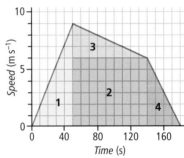

Total distance = area under graph =
(0·5 × 50 × 9) + (0·5 × 90 × 3) + (0·5 × 40 × 6) + (90 × 6) = 225 + 135 + 120 + 540 = 1020 m

(b) Calculate the average speed of the bus.

$$\bar{v} = \frac{d}{t} = \frac{1020}{180} = 5.7\,\text{m s}^{-1}$$

VELOCITY-TIME GRAPHS AND DISPLACEMENT

In a velocity–time graph, the area under the graph is equal to the displacement of a moving object.

EXAMPLE 4

This graph represents the motion of a ball thrown vertically into the air and caught again.

(a) What is the displacement of the ball after 3 s?

displacement, s = area under velocity–time graph = area of triangle
= 0·5 × 3 × 29·4 = 44·1 m (upwards)

(b) What is the displacement of the ball after 6 s?

displacement, s = area under velocity–time graph = area of triangle A + triangle B
= (0·5 × 3 × 29·4) + (0·5 × 3 × −29·4)   (Note for triangle B the velocity
= 44·1 − 44·1 = 0 m   is negative as it is *below* the
                                           time axis.)

The displacement is zero; the ball returns to exactly the same position as before it was thrown.

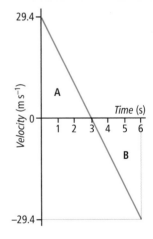

VIDEO LINK

Check out the video clip about velocity-time graphs: www.brightredbooks.net/N5Physics

THINGS TO DO AND THINK ABOUT

Motor car insurance is very expensive for young drivers. Some car insurance companies offer young drivers an app for their smart phone. This app allows the phone to monitor certain features of a car's movement and direction – its acceleration, speed, cornering and braking. Investigate how the device is able to monitor this information which is sent from the smartphone to the insurance company:

1. What use is made of this information by the insurance company?

2. How could this help the young driver?

ONLINE TEST

How well have you learned about the velocity-time graphs? Go online and test yourself at www.brightredbooks.net/N5Physics

ACCELERATION

The key concepts to learn in this topic are:

- definition of velocity in terms of initial velocity, final velocity and time
- the use of an appropriate relationship to solve problems involving acceleration, initial velocity (or speed), final velocity (or speed) and time:

$$a = \frac{v - u}{t}$$

- the determination of acceleration from a velocity–time graph:

  a = gradient of the line on a v–t graph

- description of an experiment to measure acceleration.

DON'T FORGET

Only uniform acceleration will be considered in this course.

A PRECISE DEFINITION

Acceleration is defined as the *rate of change of velocity in unit time*. For example, a car which accelerates from an initial to a final velocity has uniform (or constant) acceleration when its velocity increases by the same amount each second.

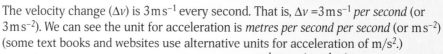

$0\,\text{ms}^{-1}$ $\Delta v = 3\,\text{ms}^{-1}$ $3\,\text{ms}^{-1}$ $\Delta v = 3\,\text{ms}^{-1}$ $6\,\text{ms}^{-1}$ $\Delta v = 3\,\text{ms}^{-1}$ $9\,\text{ms}^{-1}$

1 s   1 s   1 s

DON'T FORGET

The unit for acceleration is ms^{-2} (known as metres per second squared).

The diagram shows the velocity of a car for the first 3 seconds of its travel

The velocity change (Δv) is $3\,\text{ms}^{-1}$ every second. That is, $\Delta v = 3\,\text{ms}^{-1}$ *per second* (or $3\,\text{ms}^{-2}$). We can see the unit for acceleration is *metres per second per second* (or ms^{-2}) (some text books and websites use alternative units for acceleration of m/s^2.)

Acceleration can be determined by calculating: $\dfrac{\text{change in velocity}}{\text{time taken for change}}$

$a = \dfrac{v - u}{t}$ Where: u = initial velocity v = final velocity t = time taken

DON'T FORGET

When calculating acceleration, take care to substitute u and v correctly.

EXAMPLE 1

When overtaking a bus, a car increases its velocity from $9\,\text{ms}^{-1}$ to $21\,\text{ms}^{-1}$ in a time of $4\,\text{s}$. Calculate the acceleration of the car.

$a = \dfrac{v - u}{t}$ $u = 9, v = 21, t = 4$

$a = \dfrac{21 - 9}{4} = 3\,\text{ms}^{-2}$

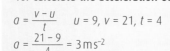

EXAMPLE 2

A cyclist travelling at $8\,\text{ms}^{-1}$ applies the brakes and comes to a rest in a time of $3.5\,\text{s}$. Calculate the acceleration of the cyclist.

$a = \dfrac{v - u}{t}$ $u = 8, v = 0, t = 3.5$

$a = \dfrac{0 - 8}{3.5} = -2.3\,\text{ms}^{-2}$ (Note the value is negative because the cyclist is decelerating.)

VIDEO LINK

For a very cool take on acceleration, check out 'Usain Bolt vs Gravity' at www.brightredbooks.net/N5Physics!

EXAMPLE 3

A typhoon jet aircraft accelerates at $7.5\,\text{ms}^{-2}$ to reach a velocity of $590\,\text{ms}^{-1}$ in $50\,\text{s}$. Calculate the initial velocity of the aircraft.

$a = \dfrac{v - u}{t}$ $a = 7.5, v = 590, t = 50$

$7.5 = \dfrac{590 - u}{50}$ so $u = 590 - (7.5 \times 50) = 215\,\text{ms}^{-1}$

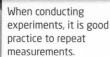
DON'T FORGET

When conducting experiments, it is good practice to repeat measurements.

EXPERIMENTS TO MEASURE ACCELERATION

A good *estimate* of the magnitude of acceleration can be obtained by measuring a small change in velocity over a small change in time.

contd

Consider a trolley with two cards of equal width attached to it. The trolley is released from the top of a slope and each card, in turn, passes through a light gate near the foot of the slope.

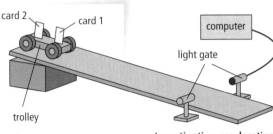
Investigating acceleration

- Times t_1 and t_2 for card 1 and card 2 to pass through the light gate are measured.
- The time t_3 – the time between card 1 and card 2 passing through the light gate is also measured.
- The initial speed u is calculated using the card width and the time t_1.
- The final speed v is calculated using the card width and the time t_2.
- The acceleration of the trolley is calculated using: $a = \dfrac{v - u}{t}$

Sample results

Card width $d = 0.05\,\text{m}$

Time for card 1, $t_1 = 0.15\,\text{s}$ $u = \dfrac{\text{card width}}{t_1} = 0.33\,\text{m s}^{-1}$

Time for card 2, $t_1 = 0.08\,\text{s}$ $v = \dfrac{\text{card width}}{t_1} = 0.63\,\text{m s}^{-1}$

Time interval $t_3 = 0.75\,\text{s}$ $a = \dfrac{v - u}{t_3} = 0.4\,\text{m s}^{-2}$

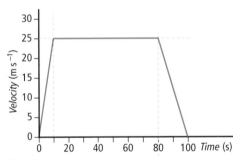
Investigating average acceleration

A similar method uses two light gates and one card length to calculate the average acceleration.

ACCELERATION FROM VELOCITY-TIME GRAPHS

The gradient of a velocity–time graph is equal to the acceleration.

The equation $a = \dfrac{v - u}{t}$ can be used to calculate the gradient of the graph and hence the acceleration.

Care is needed when selecting values for u, v, and t from the graph. For this graph,

$\text{gradient} = \dfrac{y_2 - y_1}{x_2 - x_1} = \dfrac{10 - 0}{20 - 0} = 0.5\,\text{m s}^{-2}$, so acceleration $= 0.5\,\text{m s}^{-2}$

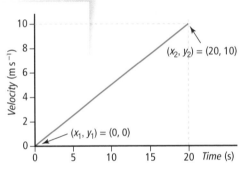

EXAMPLE 4

The graph represents the motion of a tram travelling in a straight line between two stops.

(a) Calculate the acceleration of the tram in the first 10 seconds.
Select values from the graph: $u = 0$, $v = 25$, time interval $t = 10 - 0 = 10\,\text{s}$

$a = \dfrac{v - u}{t} = \dfrac{25 - 0}{10} = 2.5\,\text{m s}^{-2}$

(b) Calculate the acceleration of the tram in the last 20 seconds.
Select values from graph: $u = 25$, $v = 0$ (tram is at rest), time interval $t = 90 - 70 = 20\,\text{s}$

$a = \dfrac{v - u}{t} = \dfrac{0 - 25}{20} = -1.25\,\text{m s}^{-2}$ (Note the negative sign, meaning deceleration and reflected in the downward slope of the graph.)

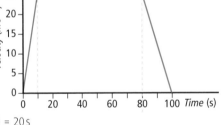

When an object is moving in a straight line and does not change direction, then its speed and velocity are *equal*. Its velocity–time graph can be labelled as either speed–time or velocity–time.

The velocity–time graph for the tram in Example 4 could have been labelled a speed–time graph, because the tram travels in one direction only. For the ball in Example 4 on p15 only a velocity–time graph is possible because the ball changes direction.

 DON'T FORGET

Choose values carefully when using velocity-time graphs.

ONLINE TEST

Go online and test yourself on 'Acceleration' at www. brightredbooks.net/ N5Physics

THINGS TO DO AND THINK ABOUT

Formula 1 racing cars must produce large acceleration. Equally, to negotiate the hairpin bends and perform overtaking measures, the cars must be capable of large deceleration. Investigate the maximum deceleration of F1 cars. Use this value to calculate the time that an F1 car would take to decelerate from $200\,\text{km h}^{-1}$ to $20\,\text{km h}^{-1}$.

NEWTON'S LAWS 1

The key concepts to learn in this topic are:

- the application of Newton's laws and balanced forces to explain constant velocity (or speed), making reference to frictional forces
- the application of Newton's laws and unbalanced forces to explain and/or determine acceleration for situations where more than one force is acting
- the use of an appropriate relationship to solve problems involving unbalanced force, mass and acceleration for situations where one or more forces are acting in one dimension or at right angles:

$F = ma$

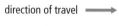

direction of travel

air resistance

less air resistance

Vehicle streamlining reduces air resistance

VIDEO LINK

Have a look at 'McLaren vs Aerodynamics' to see some modern aerodynamics in action at www.brightredbooks.net/N5Physics

NEWTON'S THREE LAWS OF MOTION

The three Newton's laws explain how forces act on objects and cause them to move or to stay still. Forces occur in many situations. A car engine provides the force which moves the car forward. Our weight is the force which keeps us on the ground.

Force of friction

Friction is a force which is present whenever an object moves or tries to move. The force of friction *opposes* the motion of an object. **Air resistance**, or drag, is a form of friction and also acts in the direction *opposite* to the motion. The frictional force of air resistance increases as the velocity of the vehicle increases.

Friction can be decreased by **streamlining** the shape of a vehicle's bodywork so the air passes over it more smoothly. The forward force of the engine can be reduced accordingly, which reduces fuel consumption and improves the fuel economy of the vehicle.

The friction on a vehicle is deliberately increased when the lorry brakes. Brake pads grip part of the wheel during braking, creating friction. When work is done against friction, kinetic energy is transformed into heat energy, thus causing the brake pads of the lorry to heat up.

NEWTON'S FIRST LAW OF MOTION: BALANCED FORCES

weight
10 000 newtons

frictional force
2000 newtons

engine force
2000 newtons

reaction force
10 000 newtons

An example of balanced forces is a car moving at constant velocity

The frictional force (2000 N) balances the engine force (2000 N)

The reaction force from the ground (10 000 N) balances the weight of the car (10 000 N)

When the forces acting on an object are *equal and opposite* we say the forces are *balanced*.

Newton's first law states that, when the forces acting on an object are balanced:

- if the object is moving, its velocity remains constant
- if the object is at rest, it will remain at rest.

Some examples of balanced forces

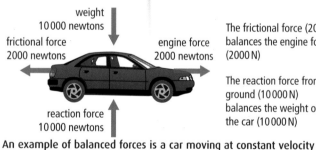

weight 10 000 N

50 km h⁻¹

frictional force
2000 N

engine force
2000 N

reaction force 10 000 N

1 balanced forces – car continues at constant velocity

weight 10 000 N

reaction force 10 000 N

2 balanced forces – zero velocity – car remains at rest

A space probe which has accelerated to extremely high velocity, perhaps 25 000 km h⁻¹,

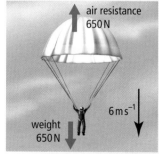

air resistance
650 N

weight
650 N

6 m s⁻¹

3 balanced forces – constant velocity

A space ship in outer space far from planets or stars experiences no air resistance

4 no forces – constant velocity

contd

will continue at this velocity when the rocket engine is switched off. There is no air to produce air resistance and no gravitational forces from planets or stars to slow it down. So, the space probe continues at constant velocity. This saves fuel and could make long distance space travel possible for future manned space exploration.

DON'T FORGET

Balanced forces mean no motion or constant velocity.

NEWTON'S SECOND LAW OF MOTION: UNBALANCED FORCES

When the forces acting on an object are *unbalanced*, the object no longer moves at a constant velocity, but will *accelerate*.

Consider the situation shown in this diagram where the forces on an object are not balanced. The resultant or unbalanced force on the object is 5 N (13 N – 8 N) to the right.

Newton's second law states: $F = ma$,
where F = the unbalanced force on the object (in N)
m = the mass of the object (in kg)
a = the acceleration of the mass (in m s^{-2})

When an unbalanced force acts on an object, its acceleration is directly proportional to the unbalanced force and inversely proportional to the object's mass.

DON'T FORGET

Unbalanced forces produce an acceleration or a deceleration.

This unbalanced force causes the object to accelerate to the right

EXAMPLE 1

Calculate the unbalanced force applied to a London bus of mass 1.7×10^4 kg when it has an acceleration of 1.21 m s^{-2}.

$F = ma = 1.7 \times 10^4 \times 1.21 = 2.06 \times 10^4$ N

EXAMPLE 2

Calculate the acceleration of a 7.5 kg object when an unbalanced force of 5.2 N is applied to it.

$a = \dfrac{F}{m} = \dfrac{5.2}{7.5} = 0.69$ m s^{-2}

EXAMPLE 3

A grasshopper exerts an unbalanced force of 0.133 N when it jumps. It accelerates at 67.5 m s^{-2}. Calculate the mass of the grasshopper.

$m = \dfrac{F}{a} = \dfrac{0.133}{67.5} = 1.97 \times 10^{-3}$ kg

ONLINE

For more extended examples go to www. brightredbooks.net/ N5Physics

unbalanced force 25 – 14 = 11 N

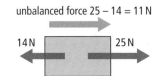

Calculating the unbalanced force

FINDING THE UNBALANCED FORCE

When using $F = ma$ it is important to remember that F *always* stands for the unbalanced force. The unbalanced force has to be determined before using $F = ma$. When more than one force acts on an object, the resultant force is the unbalanced force.

ONLINE TEST

How well have you understood Newton's laws? Go online and test yourself at www.brightredbooks.net/ N5Physics

EXAMPLE 4

A boat of mass 350 kg is sailing in a river. A current in the river exerts a force of 38 N on the boat in an east direction. A wind blowing south also exerts a force of 9 N on the boat.

(a) Calculate the resultant force acting on the boat.

(b) Calculate the acceleration of the boat.

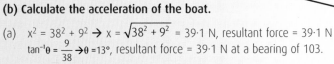

(a) $x^2 = 38^2 + 9^2 \rightarrow x = \sqrt{38^2 + 9^2} = 39.1$ N, resultant force = 39.1 N
$\tan^{-1}\theta = \dfrac{9}{38} \rightarrow \theta = 13°$, resultant force = 39.1 N at a bearing of 103.

(b) $a = \dfrac{F}{m} = \dfrac{39.1}{350} = 0.1$ m s^{-2} at a bearing of 103.

THINGS TO DO AND THINK ABOUT

Carry out research into airbags to determine how they operate to protect car occupants from injury.

NEWTON'S LAWS 2

The key concepts to learn in this topic are:

- the use of an appropriate relationship to solve problems involving weight, mass and gravitational field strength

    $W = mg$

- explanation of motion resulting from a 'reaction force' in terms of Newton's third law.

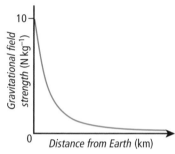

The force of gravity decreases as an object moves further away from the Earth

WEIGHT, MASS AND GRAVITATIONAL FIELD STRENGTH

The planet Earth exerts a downward force on objects. This is the gravitational force on the object and is known as the weight of the object.

The weight of an object can be calculated using the relationship:

$W = mg$ where: W = the weight of the object in newtons (N)
m = the mass of the object in kilograms (kg)
g = the gravitational field strength ($9.8\,\mathrm{N\,kg^{-1}}$ on Earth)

Gravitational field strength is defined as the *weight per unit mass* (or the force of gravity acting on each kilogram). On Earth the gravitational field strength is $9.8\,\mathrm{N\,kg^{-1}}$.

This means that each 1 kg mass on Earth experiences a gravitational, downward force of 9.8 newtons. (Other planets have different gravitational field strengths.)

The weight of an object decreases as it moves away from the Earth as shown by this graph.

The force of gravity is zero in deep space at distances well away from the Earth and other planets.

EXAMPLE 1

Calculate the weight of a gold bar of mass 12·7 kg stored in a bank vault.

$W = mg = 12\cdot7 \times 9\cdot8 = 124\cdot5\,\mathrm{N}$

EXAMPLE 2

The weight of a delivery lorry is $2\cdot45 \times 10^4$ N. Calculate the mass of the lorry.

$m = \dfrac{W}{g} = \dfrac{2\cdot45\times10^4}{9\cdot8} = 2500\,\mathrm{kg}$

EXAMPLE 3

The graph shows how the gravitational field strength varies at different heights above the surface of the Earth.

(a) What is the gravitational field strength at a height of 1750 km above the Earth?

$6\cdot0\,\mathrm{Nkg^{-1}}$

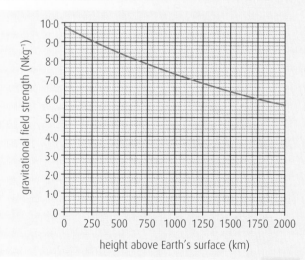

contd

(b) The International Space Station orbits above the Earth at a height of 400 km. The weight of an astronaut on board the Station is 602 N. Calculate the mass of the astronaut.

gravitational field strength at 400 km is 8·6 N kg⁻¹.

$$m = \frac{W}{g} = \frac{602}{8\cdot6} = 70 \text{ kg.}$$

NEWTON'S THIRD LAW OF MOTION

Forces exist in pairs

Newton's third law of motion states that 'If A exerts a force on B, then B exerts an *equal but opposite* force on A'.

So, Newton's third law tells us that forces exist in pairs. These Newton pairs are equal in size but opposite in direction.

Note that both forces of the pair occur at the *same time*, and are sometimes called the *action* force and the *reaction* force.

The rocket pushes the exhaust gases in one direction. The exhaust gases push the rocket in the opposite direction.

In a space rocket, the rocket exerts a downward force (called a **thrust**) on the exhaust gases; the exhaust gases exert an equal and opposite upward thrust on the rocket. This upward thrust causes the rocket to move upwards. This is why a rocket is able to accelerate in space without any atmosphere.

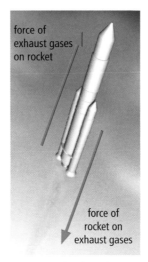

force of exhaust gases on rocket

force of rocket on exhaust gases

A space rocket launching

Examples of Newton pairs

Kicking a football: the boot exerts a forward force on the ball, the ball exerts a backwards force on the boot

A hammer striking a nail: the hammer exerts a downward force on the nail, the nail exerts an upwards force on the hammer

Pushing against a wall: your hands exert a forward force against the wall, the wall exerts a backwards force on your hands

EXAMPLE 4

In 1997, the fastest speed on land for a vehicle was recorded. The vehicle used a jet engine which pushed hot gases from its exhaust with a force of $2\cdot23 \times 10^5$ N. The mass of the vehicle was $1\cdot07 \times 10^4$ kg.

(a) On the diagram, draw the horizontal forces acting on the vehicle when it starts to move.

(b) Calculate the acceleration of the vehicle at this time.

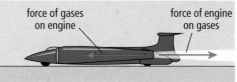

force of gases on engine

force of engine on gases

Forward force of gases = $2\cdot23 \times 10^5$ N

$$a = \frac{F}{m} = \frac{2\cdot23 \times 10^5}{1\cdot07 \times 10^4} = 20\cdot9 \text{ m s}^{-2}$$

THINGS TO DO AND THINK ABOUT

Think about the sports which interest you. For each sport, identify examples of Newton pairs of forces.

DON'T FORGET

Newton's third law is that 'for every action force there is a reaction force'.

NEWTON'S LAWS 3

The key concept to learn in this topic is:
- to explain free-fall and terminal velocity in terms of **Newton's laws**.

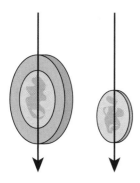

If a £2 coin and a 5p coin are dropped simultaneously from the same height they will both have an acceleration of $9.8\,\text{m}\,\text{s}^{-2}$ and hit the ground at the same time

VIDEO LINK

Watch the clip 'Feather and Coin in a Vacuum' to see this in action at www.brightredbooks.net/N5Physics

DON'T FORGET

All objects have the same acceleration when falling close to the Earth's surface, if air resistance is negligible.

VIDEO LINK

Check out the 'Speed-time graphs' clip about terminal velocity at www.brightredbooks.net/N5Physics

FREE-FALL AND TERMINAL VELOCITY

Acceleration due to gravity

The acceleration due to gravity on Earth is $9.8\,\text{m}\,\text{s}^{-2}$. When an object is dropped, the only force acting on it is its weight (discounting air resistance) – this is an unbalanced force. The object will accelerate and fall freely, with the acceleration due to gravity ($9.8\,\text{m}\,\text{s}^{-2}$). This is know as **free-fall**.

EXAMPLE 1

Show that the acceleration of an apple of mass $0.1\,\text{kg}$ is $9.8\,\text{m}\,\text{s}^{-2}$ when it falls from a tree.

Weight of apple $W = mg$
$= 0.1 \times 9.8 = 0.98$ N

Unbalanced force on falling apple
= its weight = 0.98 N

$a = \dfrac{F}{m} = \dfrac{0.98}{0.1} = 9.8\,\text{m}\,\text{s}^{-2}$

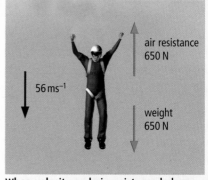

The spring balance shows the weight of the stone

Freely falling spring balance with the stone shows a zero reading

Objects in freefall appear weightless

Terminal velocity

When an unbalanced force acts on an object, it accelerates (Newton's second law). When the forces acting on a moving object are balanced, it moves at a constant velocity (Newton's first law). **Terminal velocity** occurs when the forces acting on a moving object become *balanced*.

Consider a skydiver jumping from an aircraft. The skydiver accelerates due to the downward force of his weight and his velocity increases. As the velocity *increases*, the upward air resistance force also *increases*. Eventually these two forces balance. These balanced vertical forces cause the skydiver to fall at constant velocity. This is known as terminal velocity. (For skydivers this velocity is about $56\,\text{m}\,\text{s}^{-1}$.) When the skydiver deploys the parachute, the area of the parachute makes the skydiver and parachute far less streamlined. The air resistance force increases rapidly.

This causes an upward force which decelerates the parachutist. Eventually the weight and air resistance forces balance again at a velocity of about $6\,\text{m}\,\text{s}^{-1}$. The skydiver now falls at a terminal velocity of $6\,\text{m}\,\text{s}^{-1}$, a safer velocity for landing than $56\,\text{m}\,\text{s}^{-1}$!

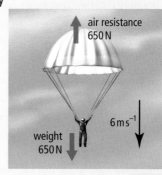

air resistance 650 N

$56\,\text{m}\,\text{s}^{-1}$

weight 650 N

When velocity and air resistance balance, the skydiver falls at constant (terminal) velocity

air resistance 650 N

weight 650 N

$6\,\text{m}\,\text{s}^{-1}$

contd

EXAMPLE 2

A ball bearing was dropped into a jar of glycerine. The velocity of the ball bearing was measured at 0·1 s time intervals, and the results recorded in the table. Use the table of results to explain the motion of the ball bearing in the glycerine.

Velocity (ms⁻¹)	0	0·20	0·40	0·60	0·80	0·95	1·04	1·06	1·12	1·12	1·12	1·12
Time (s)	0	0·1	0·2	0·3	0·4	0·5	0·6	0·7	0·8	1·0	1·1	1·2

The velocity of the ball bearing increases from zero for 0·8 s. The velocity then remains constant at 1·12 ms⁻¹. This is the terminal velocity of the ball bearing in the glycerine.

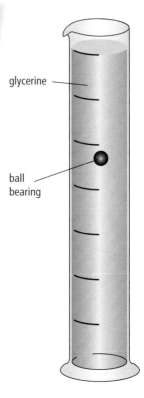

glycerine

ball bearing

EXAMPLE 3

A speedboat of mass 800 kg is initially at rest on a lake. The speedboat's engine is started and it produces a constant force of 5 kN. As the speedboat accelerates, the frictional force on it increases. A graph of the force of friction acting on the speedboat against time is shown.

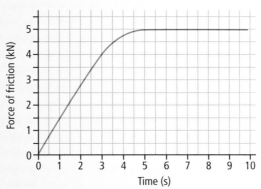

Time (s)

(a) State the force of friction acting on the speedboat 3 s after its engine is switched on.

4 kN

(b) Calculate the acceleration of the speedboat at this time.

Unbalanced force acting on speedboat = 5000 − 4000 = 1000 N

$$a = \frac{F}{m} = \frac{1000}{800} = 1·25 \text{ ms}^{-2}$$

(c) Describe and explain the movement of the speedboat after 7 s.

The engine force and frictional forces are balanced. The boat reaches its terminal velocity.

 ONLINE TEST

How well have you understood gravity, weightlessness and freefall? Go online and test yourself at www.brightredbooks.net/N5Physics

EXAMPLE 4

The velocity-time graph represents the motion of a skydiver from the instant of jumping from the aircraft until after the parachute has been opened.

Use the graph to list the following time-intervals:

(a) Which time interval on the graph shows the skydiver accelerating before the parachute is opened?

0 – 25 seconds

(b) What is the terminal velocity reached by the skydiver before the parachute is opened?

54 ms⁻¹.

(c) What is the terminal velocity reached by the skydiver after the parachute is opened?

10 ms⁻¹.

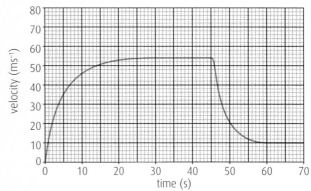

time (s)

THINGS TO DO AND THINK ABOUT

1. In October 2012, Felix Baumgartner became the first person to jump from a balloon at height of 39 km above the Earth. During his descent, Felix reached a terminal velocity which was greater than had been experienced by any other skydiver. Investigate why Felix was able to fall faster through the atmosphere than anyone had before.

ENERGY 1

The key concepts to learn in this topic are:

- explanation of energy conservation and of energy conversion and transfer
- the use of an appropriate relationship to solve problems involving work done, unbalanced force and distance/displacement:
  $E_W = Fd$, or $W = Fd$.

(a) potential energy

(b) kinetic energy

The skateboarder's potential energy transforms into kinetic energy as he rolls down the hill

ENERGY CONSERVATION AND LOSS

Energy can be transferred from one form into another, but energy cannot be destroyed. When energy is transferred, the total energy involved in the transfer is conserved. This means that the total energy before a transfer is the same as the total energy after the transfer. Sometimes energy is described as being 'lost' during a transfer, but this usually refers to the energy being transferred into a form which is not useful.

Consider a skateboarder rolling down a hill. Gravitational potential energy is transferred into kinetic energy as the skateboarder rolls down the hill.

Heat energy is produced whenever potential energy is transferred into other forms of energy – whether or not the heat energy is wanted. For example, in a car, chemical potential energy from fuel is transferred by the engine into useful kinetic energy but it is also wasted or 'lost' as heat energy.

A bouncing ball illustrates how energy is transferred.

What happens when the ball is dropped?

As the ball falls, if air resistance is ignored, then all of the gravitational E_p is transferred into E_k.

$E_p \rightarrow E_k$

What happens when the ball stops going up?

When all of the E_k is transferred back into gravitational E_p the ball will be at the top of its bounce – but <u>not</u> at the height it was dropped from because of the E_h that was lost when it changed shape.

What happens when the ball hits the ground?

As the ball hits the ground it changes shape and the E_k transfers into elastic E_p and some E_h.

What happens when the ball rebounds?

As the ball rebounds, the elastic potential energy transfers back into E_k and some E_h and it regains its shape. Then this E_k is transferred into gravitational E_p as the ball gets higher.

Energy transfers in a bouncing ball

Each time the ball bounces, the lost energy means that the ball will rebound to a lower height until eventually it loses all of its energy.

Consider a cyclist accelerating on a level road, then freewheeling up and down a small hill, and finally braking to rest. The various energy transfer are described in the diagram.

VIDEO LINK

Check out the 'Potential and Kinetic Energy' clip at www.brightredbooks.net/N5Physics

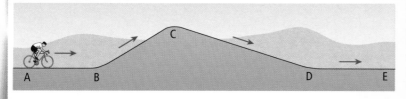

Energy transfer
A–B chemical energy to kinetic energy as cyclist accelerates from rest
B–C kinetic energy to gravitational potential energy
C–D gravitational potential energy to kinetic energy
D–E kinetic energy to heat energy as cyclist brakes and decelerates

Work done

When energy is transferred from one form into another, work is done.

Work done is another name for the amount of energy required when a force, F, is applied to move an object over a distance, d. Work is a scalar quantity.

The relationship for work is: $E_w = Fd$, (or $W = Fd$ where W is work done in joules, J)

where: E_w is the work done (in joules, J) F is the force applied (in newtons, N)
   d is the distance (in metres, m).

EXAMPLE 1

A force of 70 N moves a wheelbarrow over a distance of 30 m. Calculate the work done.

$E_w = Fd = 70 \times 30 = 2100\,\text{J}$

EXAMPLE 2

During a short journey, a metro train exerted a force of 7·5 kN and did $2·5 \times 10^6$ J of work. Calculate how far the train travelled.

$d = \dfrac{E_w}{F} = \dfrac{2·5 \times 10^6}{7·5 \times 10^3} = 333\,\text{m}$

DON'T FORGET

Work and energy have the same unit: joules.

EXAMPLE 3

A box has a weight of 65 N. It is lifted from the ground through a height of 0·5 m at a constant speed.

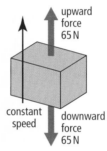

upward force 65 N

constant speed

downward force 65 N

(a) State the minimum force required to lift the box.

The minimum force is 65 N. Since the box is lifted at constant speed, the upward force must balance the weight acting downwards.

(b) Calculate the work done lifting the box if its weight is 65 N.

$E_w = Fd = 65 \times 0·5 = 32·5\,\text{J}$

VIDEO LINK

Check out the 'Energy Transfer' clip at www.brightredbooks.net/N5Physics

In many situations, when work is done, it is to overcome a frictional force. For example, when a car is moving along a road, there are frictional forces (resistive forces) opposing the motion. The car engine has to do work *against* friction, even when travelling at constant speed.

EXAMPLE 4

A 1100 kg car travelled along a 2·5 km-long road at constant speed. The car engine exerted a force of 3·5 kN.

(a) State the size of the frictional force acting on the car.

3·5 kN (Since the car moves at constant speed, the engine and friction forces are balanced.)

(b) Calculate the work done against friction.

$E_w = Fd = 3·5 \times 10^3 \times 2·5 \times 10^3 = 8·75 \times 10^6\,\text{J}$

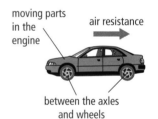

moving parts in the engine

air resistance

between the axles and wheels

Sources of friction in moving cars

THINGS TO DO AND THINK ABOUT

Car designers investigate ways to reduce frictional effects on cars when they move. If the forces of friction acting on a car as it moves can be reduced, less work needs to be done by the engine to overcome the frictional effects, requiring less energy, and so less fuel.

Carry out research to find out how the air frictional force is affected by the shape of the car, and how car designers investigate this.

ONLINE TEST

How well have you learned about work and energy? Go online and test yourself at www.brightredbooks.net/N5Physics

ENERGY 2

The key concepts to learn in this topic are:

- definition of gravitational potential energy
- use of an appropriate relationship to solve problems involving gravitational potential energy, mass, gravitational field strength and height:

$$E_P = mgh$$

- definition of kinetic energy

- use of an appropriate relationship to solve problems involving kinetic energy, mass and speed:

$$E_k = \frac{1}{2}mv^2$$

- use of appropriate relationships to solve problems involving conservation of energy:

$$E_W = Fd, \text{ or } W = Fd$$
$$E_P = mgh$$
$$E_k = \frac{1}{2}mv^2$$

GRAVITATIONAL POTENTIAL ENERGY

An object which is **at a height**, **h**, above the Earth's surface possesses gravitational potential energy (E_P).

If an object is raised upwards **to a height**, **h**, above the Earth's surface, then work is done against the force of gravity. This work is stored as gravitational potential energy.

The relationship used to calculate gravitational potential energy is:

$$E_P = mgh$$

where E_P is the gravitational potential energy (in Joules, J)

       m   is the mass of the object (in kilograms, kg)

m

h

g   is the gravitational field strength (9.8 Nkg^{-1} on Earth)

h   is the height above the surface of the Earth (metres, m)

EXAMPLE 1

Calculate the gravitational potential energy of a cat of mass 1·8 kg which is sitting on a window ledge 2·6 m above the ground.

$E_P = mgh = 1.8 \times 9.8 \times 2.6 = 45.9 \text{ J}$

EXAMPLE 2

A kite of mass 0·3 kg is caught in a tree. The kite has 10·8 J of gravitational potential energy. Calculate the height of the kite above the ground.

$E_P = mgh, h = \dfrac{10.8}{0.3 \times 9.8}, h = 3.7 \text{m}$

KINETIC ENERGY

Kinetic energy (E_k) is the energy possessed by moving objects.

The kinetic energy of a moving object depends on its speed, v, and mass, m.

The relationship used to calculate kinetic energy is

$$E_k = \frac{1}{2}mv^2$$

where E_k is the kinetic energy (in joules, J)

       m   is the mass of the object (in kilograms, kg)

       v   is the speed of the object (in metres per second, ms^{-1})

EXAMPLE 1

A car of mass 800 kg is moving at a speed of 25 ms^{-1}. Calculate the kinetic energy of the car.

$E_k = \frac{1}{2}mv^2, Ek = \frac{1}{2} \times 800 \times 25^2 = 250\ 000 \text{ J}$

EXAMPLE 2

An arrow of mass 0·2 kg is fired from a bow towards a target.
Calculate the arrow's speed if it has 1440 J of kinetic energy as it leaves the bow.

$E_k = \dfrac{1}{2}mv^2, v = \sqrt{\dfrac{2 \times 1440}{0.2}} = 120 \text{ ms}^{-1}$

CONSERVATION OF ENERGY

When energy is transferred from one type into another, energy is conserved. This means that the total energy before and after the transfer is the same. This is known as conservation of energy.

Some examples of conservation of energy are:

A car moving along a road   (chemical energy in fuel $\rightarrow E_K$)

A kettle boiling   (electrical energy $\rightarrow$ heat energy)

A ball bouncing   ($E_P \rightarrow E_K$ (downward motion), $E_K \rightarrow E_P$ (upward motion))

In these examples, some energy is transferred into 'lost' energy, which is not useful.

For example, when a car engine uses fuel, some heat energy is produced. In the kettle boiling, some sound energy is produced. In the bouncing ball, some heat energy is produced on each bounce.

EXAMPLE:

This example illustrates how to calculate speeds, height and lost energy.

(a) A 600 g ball is dropped to the ground from a vertical height of 4 m. Calculate the speed of the ball just as it collides with the ground. Ignore air resistance.

All of the energy is transformed from potential energy into kinetic energy as the ball falls because air resistance is ignored.

$$E_p \rightarrow E_k \quad \text{and} \quad mgh = \frac{1}{2}mv^2$$

$$\text{so } v = \sqrt{2gh} = \sqrt{2 \times 9 \cdot 8 \times 4} = 8 \cdot 9 \text{ ms}^{-1}$$

(b) When the ball rebounds, it leaves the ground with a speed of 7·4 ms⁻¹. Calculate the height the ball will return to.

This time E_k is transformed into E_p. So $v = \sqrt{2gh}$ rearranges to $h = \frac{v^2}{2g}$

$$h = \frac{7 \cdot 4^2}{2 \times 9 \cdot 8}$$

$$h = 2 \cdot 8 \text{ m}$$

(c) Calculate the energy lost when the ball collides with the ground.

Before the collision $E_k = \frac{1}{2}mv^2 = \frac{1}{2} \times 0 \cdot 6 \times 8 \cdot 9^2$

$$= 23 \cdot 8 \text{ J}$$

After the collision $E_k = \frac{1}{2}mv^2 = \frac{1}{2} \times 0 \cdot 6 \times 7 \cdot 4^2$

$$= 16 \cdot 4 \text{ J}$$

Energy lost during collision   = E_k before – E_k after collision

$$= 23 \cdot 8 - 16 \cdot 4$$

$$= 7 \cdot 4 \text{ J}$$

THINGS TO DO AND THINK ABOUT

Here are some other examples of everyday energy transformations:

Water that is stored in a reservoir and then flows downwards through pipes to a generator in a hydroelectric power station

The movement of a clock pendulum

1. Think of some examples of energy transfers that are going on around you.

2. Consider the energy losses which may occur in these transfers. Why are processes less than 100% efficient in terms of useful energy?

PROJECTILE MOTION

The key concepts to learn in this topic are:

- explanation of projectile motion in terms of constant vertical acceleration and constant horizontal velocity
- explanation of satellite orbits in terms of projectile motion, horizontal velocity and weight
- the use of appropriate relationships to solve

problems involving projectile motion from a horizontal launch, including the use of motion graphs:

area under v_h-t graphs (horizontal range)

area under v_v-t graphs (vertical height)

$v_h = \dfrac{s}{t}$ (constant horizontal velocity)

$v_v = u_v + at$ (constant vertical acceleration)

PROJECTILES

DON'T FORGET

On this course, only projectiles which have been launched horizontally will be considered.

A **projectile** is any object that has been launched (or projected) into the air. Once in the air, gravity is the only force acting on the projectile (assuming air resistance is negligible).

Horizontally-launched projectiles move outwards away from the launch site as well as falling vertically, and so they move in curved paths.

The projectile's curved path is caused in part by the force of gravity – the object accelerates downwards. At the same time the horizontal velocity stays constant if air resistance is negligible.

Compare one object dropped vertically with another identical object projected horizontally at the same time.

Both objects reach the ground at exactly the *same time*. This result shows that the time of flight for the projectile depends *only* on its initial vertical height, not on its horizontal motion.

Projectiles travelling through the air

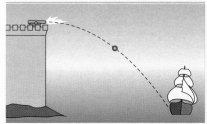

The cannon ball shown here is an example of a projectile that has been launched horizontally

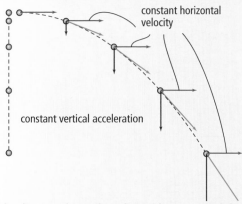

constant horizontal velocity

constant vertical acceleration

The diagram shows that after each time period the vertical positions of the objects match

PROJECTILE MOTION

DON'T FORGET

The time of flight for a projectile is the same for horizontal and vertical motion.

Projectile motion can be treated as two independent motions. The horizontal motion can be treated separately from the vertical motion when solving numerical examples.

- The horizontal velocity of a projectile stays constant as there are no horizontal forces acting on the projectile (air resistance is negligible).
- The vertical velocity of a projectile increases as the projectile accelerates under the force of gravity, with an acceleration of $9.8\,\text{m s}^{-2}$.

Graphs can be drawn to represent the horizontal and vertical motions as shown in the diagram.

The time of flight for the projectile is the same for the horizontal and vertical motion.

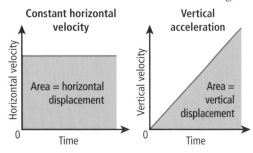

Graphs for the horizontal and vertical components of motion

Constant horizontal velocity

Horizontal velocity

Area = horizontal displacement

0   Time

Vertical acceleration

Vertical velocity

Area = vertical displacement

0   Time

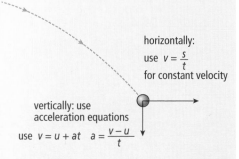

horizontally:
use $v = \dfrac{s}{t}$
for constant velocity

vertically: use acceleration equations

use $v = u + at$ $a = \dfrac{v-u}{t}$

contd

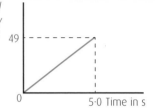

EXAMPLE 1:

An aircraft flying horizontally at $55\,\text{m s}^{-1}$ drops a package of supplies to earthquake victims. The package has a vertical velocity of $75\,\text{m s}^{-1}$ when it hits the ground.

(a) Calculate the time taken for the package to reach the ground.

$a = \dfrac{v-u}{t}$ for vertical motion, $u = 0$, $v = 75$,

$a = 9.8$

$9.8 = \dfrac{75-0}{t}$ so $t = \dfrac{75-0}{9.8} = 7.65\,\text{s}$

(b) Calculate the horizontal distance travelled by the package before it reached the ground.

The package is released horizontally, so $v_h = 55\,\text{m s}^{-1}$

$s = vt = 55 \times 7.65 = 421\,\text{m}$

(c) State any assumptions made when carrying out your calculations.

Assume that air resistance has a negligible effect on the package's movement.

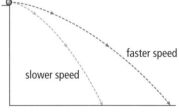

ONLINE TEST

Go online and test yourself at www.brightredbooks.net/N5Physics

EXAMPLE 2:

A cannon ball is fired horizontally from a cliff as part of a battle re-enactment.

The following graphs represent the horizontal and vertical motions of the cannonball as it leaves the cannon until it reaches the ground. The effect of air resistance is negligible.

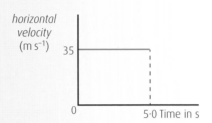

(a) Explain why the path of the cannonball is curved.

The cannonball moves horizontally at constant velocity, and at the same time falls vertically with constant acceleration.

(b) What is the horizontal velocity of the cannonball as it leaves the cannon?

$35\,\text{m s}^{-1}$

(c) Determine the height above the cliff at which the cannonball was fired.

Vertical height = area under vertical velocity-time graph,

height $= \dfrac{1}{2} \times 5.0 \times 49 = 122.5\,\text{m}$

(d) A second cannonball is fired from the cannon which lands further from the cliff.

Calculate the velocity of the cannonball as it leaves the cannon.

Time of flight is still 5.0 seconds because vertical height is same.

horizontal velocity $v = \dfrac{s}{t} = \dfrac{280}{5} = 56\,\text{m s}^{-1}$

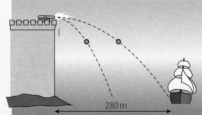

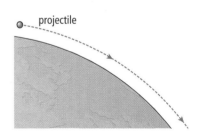

The projectile that is fired at the fastest velocity lands the furthest from the edge of the cliff

A projectile with a curved path that matches the curvature of the Earth

SATELLITE MOTION

Satellite motion is an extension of projectile motion. The faster a projectile is fired horizontally, the greater the distance it travels before reaching the Earth's surface. The diagram shows a projectile being fired horizontally off a cliff at two different velocities.

The surface of the Earth is flat on this diagram, but the Earth is a sphere so its surface is curved. If a projectile has a fast enough initial horizontal velocity, the shape of its curved path will be the same as the shape of the Earth's curved surface.

The projectile is falling towards the Earth but never reaches the surface – the curved projectile path and the Earth's curvature are the same. An artificial satellite is put into orbit by launching it from Earth, then giving it the correct horizontal velocity so that it falls back to Earth with a curved path that matches the Earth's curvature. The satellite motion takes place above the Earth's atmosphere so there is no friction to slow down the satellite.

A satellite with a curved path around the Earth

SPACE

SPACE EXPLORATION 1

The key concepts to learn in this topic are:
- basic awareness of our current understanding of the universe
- use of the following terms correctly and in context: planet, dwarf planet, moon, Sun, asteroid, solar system, star, exoplanet, galaxy, universe.

THE UNIVERSE

The universe is a term that is used to describe everything that exists in the entirety of space.

The current understanding of the universe is of a collection of countless galaxies. Some astronomers' estimates mention a figure of 2 trillion galaxies (2×10^{12} galaxies). These estimates frequently change, as more information is obtained about the nature of the universe. Also, as more information becomes available, scientists refine their theories about the origins and structure of the universe.

Current theories regarding the origin of the universe favour the 'Big Bang' model. This describes an 'event' happening about 14 billion years ago leading to the continuing formation of stars and galaxies.

ASTRONOMICAL TERMS

Our universe contains many different objects. As new information from space exploration is received, the list of objects that have been discovered, or are thought to theoretically exist, continues to grow.

Some of the astronomical terms used to describe objects in our universe are given in the table.

Term	Definition	Facts
Planet	A natural satellite that orbits a star	Eight planets orbit our local star (the Sun). Note: Pluto is now described as a dwarf planet
Dwarf planet	A dwarf planet is an object which orbits a star, has a mass which is sufficient to produce a round-type planet, but shares its own orbit with other objects such as asteroids or comets	Other dwarf planets in addition to Pluto have been discovered in the solar system, such as Cers, Eris, Haumea and Makemake, and there may be many more as yet undiscovered
Asteroid	Asteroids are solid objects in space, smaller than planets, orbiting the sun	Most asteroids are found in the space between the orbits of the planets Mars and Jupiter
Exoplanet	A planet outside our solar system, orbiting another star	To date, more than 2000 exoplanets have been discovered and this number is increasing all the time (also known as extrasolar planets)
Moon	A natural satellite that orbits a planet	Astronomers have discovered at least 140 moons orbiting other planets in our solar system
Star	A mass of hot gas that emits heat and light	Vast amounts of energy are produced in the cores of stars. Enormous temperatures and pressures cause nuclear fusion - nuclei of hydrogen join together to form helium and produce energy in the process
Solar system	A star and the objects that orbit it	Planets, moons and asteroids (smaller lumps of rock) orbit our Sun. Comets also orbit the Sun, but usually have a long orbit cycle
Sun	The star at the centre of our solar system	It is estimated that the Sun is approximately 4·6 billion years old and that it is halfway through its life
Galaxy	A collection of stars	Astronomers have identified three main types of galaxy: spiral, elliptical and irregular
Universe	All the energy and matter (galaxies, stars, planets, interstellar gas and space) in existence	The current theory for the formation of the universe is the Big Bang model. The Big Bang model estimates the age of the universe to be approximately 13·7 billion years.
Dark matter	Scientists believe, from theoretical calculations, that there is a huge amount of matter in the universe that is so far undetected by instruments	The behaviour of the observable universe indicates that there may be undetected matter - referred to as 'dark matter'
Dark energy	Scientists believe that there is a source of energy that keeps the universe expanding, but they have not been able to identify it	Scientific instruments have so far been unable to detect dark energy

Scale of the solar system and universe measured in light years

The diameter of our solar system measured across the outermost planetary diameter is approximately 0·0013 light years or $1·23 \times 10^{13}$ m.

The Earth is approximately 46·5 billion light years from the most distant edge of the observable universe.

The Andromeda galaxy

EXAMPLE

The table gives information about some of the planets in our solar system.

Planet	Mercury	Venus	Earth	Mars	Jupiter	Saturn	Neptune
Distance from the Sun (million km)	58	110	150	228	780	1430	4500
Time to orbit the Sun once (Earth-years)	0·24	0·6	1	1·9	12	29	165
Time for one complete spin (in Earth-days or Earth-hours)	59 days	243 days	24 hours	25 hours	10 hours	10 hours	16 hours
Gravitational field strength (N kg⁻¹)	3·7	8·9	9·8	3·7	23	9·0	11

The solar system

(i) **Which two planets have the same length of day?**
Jupiter and Saturn

(ii) **On which planets will a 4 kg mass have the same weight?**
Mercury and Mars

(iii) **Which planet has the longest solar orbit time?**
Neptune

Exoplanets

Interest in exoplanets has increased since the first discovery of an exoplanet orbiting a distant star. Scientists think that some exoplanets may have suitable conditions that might be able to support life in some form.

Scientists refer to the **habitable zone** around a star. This is the zone in which a planet would have the appropriate conditions (such as surface pressure and temperature) to allow liquid water to exist. Water is essential to all life on Earth. The Earth is used here as a model to describe the requirements for life to exist on other planets.

The table gives the basic requirements for a planet to be in a potentially habitable zone around a star.

Requirement	Additional information
Liquid water	Wide expanses of water in liquid form should be available
Suitable temperature	The planet would need a surface temperature within the limits required for life to exist (i.e. no huge extremes in surface temperatures)
Oxygen	This is required by humans and most other species on Earth
Food	The planet would need conditions that allowed food to be produced

The enormous distances to other stars that support exoplanets means that travel to an exoplanet is not currently possible. The fastest speed possible in the universe is the speed of light in a vacuum (3×10^8 m s⁻¹). Even if a spacecraft could travel close to this speed, it would take years to reach the planet. Current space travel technology cannot meet these requirements.

DON'T FORGET

Be able to describe the different types of object in the universe.

Distant galaxies

THINGS TO DO AND THINK ABOUT

Several unmanned space probes have been launched into our solar system to study planets and their moons.

Investigate some of the most recent probes which have been launched. Construct a table to record the name of the probe, its mission, and what information has been sent back to Earth. Use reliable websites, like NASA for data.

SPACE EXPLORATION 2

The key concepts to learn in this topic are:
- awareness of the benefits of satellites: GPS, weather forecasting, communications, scientific discovery and space exploration (for example Hubble telescope, ISS)
- knowledge that geostationary satellites have a period of 24 hours and orbit at an altitude of 36 000 km
- knowledge that the period of a satellite in a high altitude orbit is greater than the period of a satellite in a lower altitude orbit.

Satellite uses

THE BENEFITS OF SATELLITES

Application satellites have had an overwhelming effect on day-to-day living for mankind. These are satellites which have been placed in orbit for a specific purpose.

For example, worldwide communication and data transfer – this has improved communications, and has had an impact on the everyday lives of many people.

Radio waves cannot travel directly between distant receivers on the surface of the Earth because of the curvature of the Earth's surface. Instead, the signals are transmitted to a satellite, which then transmits this signal to a receiving Earth station. This has increased the amount and speed of communication between countries.

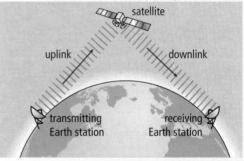

Satellites are used to transmit signals over the curved surface of the Earth

The radio signals used to communicate with satellites travel at the speed of light (3×10^8 m s^{-1}). Satellites are a significant distance from the Earth's surface. This means that there is a time delay between sending and receiving a signal.

EXAMPLE

A global positioning system communications satellite is in orbit 20 000 km above the Earth.

Calculate the time taken for a signal sent from Earth to reach the satellite.

$$20\ 000\ \text{km} = 20\ 000 \times 1000\ \text{m} = 2 \times 10^7\ \text{m}$$
$$t = \frac{d}{v} = \frac{2 \times 10^7}{3 \times 10^3} = 0{\cdot}067\ \text{s}$$

A signal sent from a weather satellite takes 0·12 s to reach a receiver on Earth.

Calculate the distance of the satellite from the receiver.

$$d = vt = 3 \times 10^3 \times 0{\cdot}12 = 3{\cdot}6 \times 10^7\ \text{m (36 000 km)}$$

Further benefits produced by satellites include:
- global and local weather information and forecasting, including tracking of hurricanes and other storm systems
- monitoring of long-term climate change – this has highlighted the importance of reducing activities which lead to global warming
- pollution monitoring – this has identified areas of the planet where people are suffering from the effects of pollution (for example, some parts of the world are affected by exposure to increased ultraviolet radiation resulting from deterioration of the ozone layer by pollution)
- provision of Global Positioning Systems – satellite navigation is commonly used today
- military observations
- commercial entertainment – satellite broadcasting television channels now mean that events can be viewed worldwide almost as they happen.

Thermometer that measures infrared radiation emitted from the body

contd

Scientists and engineers develop technology and equipment for space exploration for use in satellites. Much of this technology has been modified for use in everyday life.

For example, some of the medical benefits which developed from this space research include:

- Space research and development into instruments for use onboard satellites to measure the amount of infrared radiation emitted from distant stars and planets. This expertise was used to produce infrared thermometers which determine body temperature accurately by measuring the infrared radiation emitted from the inner ear, without actually being in contact with a patient. This has greatly reduced the instances of cross infection.

- Technology which was developed to improve images of the moon has been adapted for use with body imaging scanning devices.

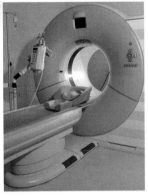

MRI scanners developed from technology that was used to observe the moon

SATELLITE ORBITS

Satellites are placed into orbit around the Earth and are used for many different purposes.

Satellites are used for communication, to gather data about the Earth (e.g. for weather forecasting), for military observations and information-gathering, and even for the observation of electromagnetic radiation from space.

Height of orbit

The height of a satellite's orbit above the Earth's surface depends on its intended use. Satellites in low orbits must travel faster than satellites in higher orbits. The International Space Station orbits the Earth once every 90 minutes and thus orbits the Earth 16 times each day.

The time taken for a satellite to go once round the Earth depends on the height of the orbit above the Earth. The period, T, of the satellite is the time taken for it to complete one rotation around the Earth. The further away the satellite is from the Earth, the longer the period.

Geostationary satellites

A **geostationary satellite** orbits at a height where a complete orbit of the Earth takes 24 hours. Therefore, as the satellite orbits it remains above the same point on the Earth's surface as the Earth rotates. The height that allows a satellite to make one orbit in 24 hours is an altitude of approximately 36 000 km above the Earth's surface. Communications and weather satellites are often placed in geostationary orbits because their data then always relates to the same area on the Earth's surface.

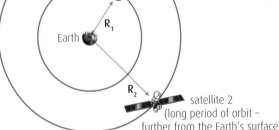

R_1 (radius of satellite 1)
R_2 (radius of satellite 2)

satellite 1
(short period of orbit – closer to the Earth's surface)

Earth

R_1

R_2

satellite 2
(long period of orbit – further from the Earth's surface)

The further away from the Earth, the longer the period of the satellite

The time taken to orbit the Earth depends on the height of the satellite

THINGS TO DO AND THINK ABOUT

Satellite orbits are usually separated into three classes depending on their altitude: Low Earth Orbit (LEO), Mid Earth Orbit (MEO) and High Earth Orbit (HEO).

1. Carry out research into satellites that are in LEO, MEO and HEO orbits.

2. Group together satellites that have common uses – for example weather forecasting, and global positioning satellites.

3. Use reliable search engines to find out about the sensors that some satellites use to obtain information, when the satellite was launched, how it obtains energy and its lifespan.

SPACE EXPLORATION 3

The key concept to learn in this topic is:

● awareness of the challenges of space travel:

— travelling large distances with the possible solution of attaining high velocity by using ion drive (producing a small unbalanced force over an extended period of time)

— travelling large distances using a 'catapult' from a fast moving asteroid, moon or planet

— manoeuvring a spacecraft in a zero friction environment, possibly to dock with the ISS

— maintaining sufficient energy to operate life support systems in a spacecraft, with the possible solution of using solar cells with area that varies with distance from the Sun.

ONLINE

Read more about ion drive engines at www.brightredbooks.net/N5Physics

DON'T FORGET

Space exploration has helped our understanding of the universe, and has produced many benefits for people.

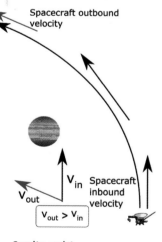

Gravity assist

THE CHALLENGES OF SPACE TRAVEL

Travelling large distances in space.

Rocket engines use a huge amount of conventional fuel at launch simply to propel a spacecraft into space. These rocket engines are discarded after the launch into space. The spacecraft itself has a limited capacity for carrying conventional fuel.

Space exploration of other planets requires vast distances to be travelled. Once launched, spacecraft have to be able to travel at high speeds to reach distant destinations in a reasonable time. Possible solutions for travelling large distance in space have already been developed:

Ion drive engine

An ion drive engine produces a beam of gas ions. When these ions are expelled with a force (thrust) from the engine nozzle, there is a reaction force on the nozzle. This reaction force causes acceleration of the spacecraft. This force is typically very small, and, using $a = \frac{F}{m}$, means that the acceleration is very small. The ion thrust engine must operate for a long time before the spacecraft reaches its top speed. The electrical energy required to operate the ion drive is obtained from a solar panel array or a nuclear generator.

NASA's Dawn Probe was launched in 2007 on a journey to the distant asteroid belt which lies between Mars and Jupiter, to study the asteroid Vesta and dwarf planet Ceres. The Probe, which uses an ion drive propulsion system powered by a solar panel array, reached Vesta in July 2011, and Ceres in March 2015.

Gravity assist

Gravity assist (sometimes known as a 'catapult' of 'gravitational slingshot') is a technique which allows a spacecraft to increase its speed by 'flying by' a larger object in space such as a fast-moving asteroid, moon or planet. For example, as the spacecraft approaches a planet, a gravitational attractive force will cause the spacecraft to start orbiting the planet. The planet is moving in its own orbit around the sun. When the spacecraft leaves its orbit around the planet some of the kinetic energy of the planet is transferred to the spacecraft and its speed increases. On its mission to study the dwarf planet Pluto, the NASA spacecraft New Horizons used this gravity assist technique by flying close to the planet Jupiter, to increase its speed and so shorten its overall journey time to Pluto.

Manoeuvring a spacecraft in orbit

Soyuz rocket vehicles are used to launch manned and unmanned Soyuz spacecraft into orbit.

Soyuz spacecraft are used to provide services for the International Space Station (ISS). This includes ferrying crew to and from the Station, supplying essentials like food and equipment, and apparatus to conduct experiments in space conditions.

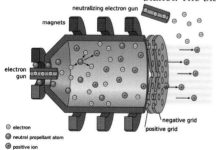

Ion drive engine

contd

When it reaches the ISS in orbit above the Earth, the Soyuz spacecraft has to connect to or 'dock' with the Space Station.

The spacecraft contains equipment which allows the docking process to safely take place:

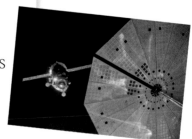

- a periscope to allow the crew to view the docking target as they approach the ISS
- a guidance system to assist the docking manoeuvre
- propulsion thrusters which fire jets of gas to adjust the speed and direction as the spacecraft approaches the ISS.

In space orbit around the Earth or other planets, there is very little friction because of the reduced atmosphere. Thrusters on the spacecraft expel exhaust gases with a small force. The exhaust gases exert an equal and opposite reaction force on the thrusters, according to Newton's third law, which causes the spacecraft to move in the opposite direction.

As the spacecraft approaches the ISS, minor adjustments of its speed and position are required to safely dock with the ISS. This is usually done using the automated guidance system on board, but can be conducted manually by the crew when necessary.

MAINTAINING SUFFICIENT ENERGY TO OPERATE LIFE SUPPORT SYSTEMS

During a long manned voyage in space, electrical energy is required to operate equipment in the spacecraft to maintain a suitable temperature, produce oxygen and water and to remove waste gases from the air. This electrical energy is also used to recharge batteries.

Solar (photovoltaic) cells are used to transform solar into electrical energy, and usually recharge batteries for use when the panels are shaded from the sun. As a spacecraft moves further away from the sun, the solar energy falling on the cells reduces. Using current technology, solar cells are practical for spacecraft operating at a distance up to the orbit of Mars. One suggestion is to increase the size of the solar cell array as the spacecraft travels further away from the sun. This would help maintain the level of solar energy received by the spacecraft on its journey through space.

The Rosetta spacecraft with extended solar panel arrays

Radioisotope thermoelectric generators

(RTGs) use radioactive material which produces heat energy when it decays. The heat energy is converted into electrical energy by devices called thermocouples.

The radioactive source Plutonium-238 is commonly used as the source radioactive element in the RTG. Plutonium-238 primarily emits alpha radiation so requires little shielding. It has a long half-life of 88 years which ensures a prolonged heat energy output during long space missions. The size of the fuel package is also compact, reducing the overall mass required to fuel the RTG.

The curiosity space vehicle on Mars uses an RTG as an energy source

Currently, the typical electrical power output from this type of generator is quite low; however several could be combined to provide greater power.

For space exploration missions beyond the planet Mars, light energy from the sun is too low to power solar panels on a spacecraft. RTGs have been used as a power source in many NASA space exploration missions to distant planets and beyond.

ONLINE TEST

How well have you learned about space exploration? Test yourself at www.brightredbooks.net/N5Physics

THINGS TO DO AND THINK ABOUT

RTGs are a source of power for spacecraft on distant missions away from the sun. Research the space missions to different parts of the solar system and beyond which have used RTGs to provide electrical energy to maintain operation of the spacecraft, and communication systems with Earth.

Make a list of the missions and indicate which spacecraft are still operating in space.

SPACE EXPLORATION 4

The key concepts to learn in this topic are:
- awareness of the risks associated with manned space exploration:
  - fuel load on take-off
  - potential exposure to radiation
  - pressure differential
  - re-entry through an atmosphere

NASA's Launch Abort System (LAS)

ONLINE

Check out NASA's Orion program at www.brightredbooks.net

NASA's Orion Space capsule's return to Earth

The Space Shuttle looked much like a conventional aircraft on re-entry

ONLINE TEST

How well have you learned about this topic? Take the 'Space Exploration' test at www.brightredbooks.net/N5Physics

THE RISKS ASSOCIATED WITH MANNED SPACE EXPLORATION

Launching a space rocket: fuel load on take off

There are many difficulties connected with the launching of a space vehicle. Vast amounts of fuel have to be carried on the rocket to maintain the thrust of the rocket's engines as it accelerates into space. This means that take-off is a very dangerous time for the astronauts. Once a rocket has been launched, the astronauts are at risk until they have been safely delivered in their spacecraft into orbit.

NASA is currently developing space rocket technology to take astronauts further into space beyond an Earth orbit, with a safe return. An escape vehicle (Launch Abort System) has been developed which will pull astronauts away from a falling rocket, and allow astronauts to land safely in the ocean using parachutes, if a problem arises at take-off.

Potential exposure to radiation in space

Radiation in space consists mainly of high energy ionising particles which could harm humans if they are not protected. The main sources of these particles are

- cosmic rays originating from deep space
- solar flares
- ionising particles which are held within regions of the magnetic field surrounding the Earth.

On board instruments constantly monitor radiation levels during spaceflights. Beyond the Earth's magnetic field, radiation levels can become potentially lethal. Research is constantly being carried out to find ways of minimising the risks of over exposure to crew members if more distant space manned exploration is undertaken.

Pressure differential: withstanding the pressure differences in space

When a spacecraft leave the Earth's atmosphere, the air pressure outside the spacecraft reduces. In space, the pressure is reduced to near vacuum levels. The cabin pressure must be maintained to allow the occupants to survive. This means that the spacecraft must be able to withstand a huge pressure difference between the inside and the outside of the cabin. This also requires a complex system to monitor and maintain life support systems to maintain the oxygen, temperature and pressure levels necessary for survival.

Re-entry to a planet's atmosphere

Space vehicles and astronauts have to survive the descent through the Earth's (or other planet's) atmosphere on their return.

When returning from orbit or space, the spacecraft has a huge amount of kinetic energy. As the spacecraft gets closer to the Earth's surface, its speed and kinetic energy must be reduced.

This kinetic energy is gradually transferred into heat energy as the spacecraft leaves orbit and enters the Earth's atmosphere. The return of the spacecraft has to be controlled as it returns from its orbit. If the angle of descent is too shallow, the spacecraft would bounce off the atmosphere and return into orbit and, if its descent is too steep, too much heat energy would be produced in a short time, destroying the craft. Passengers and cargo must be protected from extremely high temperatures. Returning spacecraft are fitted with a 'heat shield' material to insulate and protect occupants from excessive temperatures.

There are two methods of descent for the dissipation of heat energy during re-entry:

contd

1. Re-entry using *flight descent* – for large spacecraft, a huge amount of kinetic energy has to be transferred into heat energy. A lengthy time of descent is required so that the rate of heating is reduced to a manageable level. The spacecraft has to be able to fly (or glide since it has no engines) through the atmosphere to descend at this lower rate. The Space Shuttle was an example of a spacecraft which had a controlled flight descent. When it landed, it did so as a conventional aircraft on a long runway. Flight descent to Earth requires the spacecraft structure to have wings. The protruding wings cause problems with heating at the start of re-entry.

2. Re-entry using *ballistic descent* – this type of descent is when the spacecraft is steered into the atmosphere to return, almost in freefall, directly to the surface. Ballistic re-entry takes a short time (typically less than 1 hour). One method of reducing the kinetic energy of the returning vehicle is to separate a 'descent module' or 'entry module' with a smaller mass from the rest of the spacecraft before leaving orbit. This is known as 'mass shedding'. Only this descent module, carrying the astronauts, returns to Earth. One disadvantage of this method is that the size of the descent module is also limited. The 'descent module', used in this type of re-entry in the Soyuz Spacecraft programme, is designed with a specific shape to protect the interior from overheating. Once the descent has started, there is limited control which can be exerted on the craft, except to deploy parachutes when it is closer to the Earth's surface. The module also has two sets of three small engines on the bottom of the vehicle which fire, one second before touchdown, slowing the vehicle to soften the landing. Much of the mass of the returning descent module consists of its heat shield, which protects the occupants or cargo from the heat produced around the descent module on re-entry.

Launch of a Soyuz space rocket carrying the capsule that transports scientists to the International Space Station.

A spacecraft's heat shield has several features to help reduce the heating of the interior.

Some heat shields use a complex process known as *ablation*. One ablation process during re-entry involves disintegration of the surface material of the heat shield at extreme temperatures, removing heat energy as the material leaves the shield. Another ablation process occurs at high temperature when some of the material inside the heat shield changes state into gas, absorbing **latent heat energy** of vaporisation. The pressure of the gas forces it out of the heat shield, removing more heat energy. The escaping gases also prevent external hot gases from reaching the module. The Soyuz descent module, and NASA's Orion spacecraft uses a heat shield which ablates.

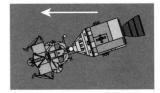

A descent module detaching from the parent spacecraft.

Some heat shields use a process known as *dissipation*. The heat energy is absorbed by insulating tiles which cover the spacecraft. Some of the absorbed heat energy is then re-radiated from the spacecraft back into the atmosphere, to maintain an acceptable temperature level inside the manned compartment. The Space Shuttle used dissipation.

EXAMPLE

The Soyuz descent module, including astronauts, has a mass of 2900 kg. During its return through the Earth's atmosphere, its drogue and main parachutes decelerate the module from 230 ms⁻¹ to 7 ms⁻¹ in 900 s.

(i) Calculate the weight of the module
$$W = mg = 2900 \times 9.8 = 2.8 \times 10^4 \text{ N}$$

(ii) Calculate the deceleration of the module.
$$a = \frac{v-u}{t} = \frac{7-230}{900} = -0.25 \text{ ms}^{-2}$$

(iii) Calculate the average unbalanced force causing this deceleration.
$$F = ma = 2900 \times -0.25 = -725 \text{ N}$$
(this is an upward force)

(iv) Calculate average the upward force exerted by the parachutes on the module.

upward force = weight + unbalanced upward force

upward force = $2.8 \times 10^4 + 725 = 2.9 \times 10^4$ N

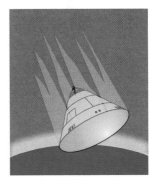

Some heat shields use ablation to remove heat on re-entry.

 DON'T FORGET

You need to know about the problems associated with re-entry into the Earth's atmosphere.

 THINGS TO DO AND THINK ABOUT

Carry out research into the risks and challenges associated with manned space exploration. Visit reputable websites like NASA and ESA (the European Space Agency).

SPACE EXPLORATION 5

The key concepts to learn in this topic are:

- knowledge of Newton's second and third laws and their application to space travel, rocket launching and landing
- use of an appropriate relationship to solve problems involving weight, mass and gravitational field strength, in different locations in the universe: $W = mg$.

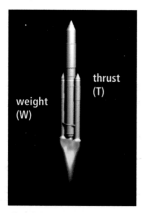

thrust (T)

weight (W)

The thrust is greater than the weight – this results in an unbalanced upward force which causes the rocket to accelerate upwards

ROCKET LAUNCHING AND LANDING

When a rocket launches to leave or lands to touch down on a planet's surface, unbalanced forces are required to accelerate or decelerate the rocket.

Rocket launching

Newton's third law accounts for how a spacecraft can accelerate at lift-off from a planet, and in space in a vacuum.

At lift-off, there are two forces acting on the rocket: the rocket's weight downward and the upward force (or thrust) from the exhaust gases. The rocket's engines exert a downward force on the exhaust gases, the exhaust gases exert an equal and opposite upward thrust on the rocket's engines. If the upward thrust from the exhaust gases is greater than its weight, then the rocket will accelerate. If the upward thrust from the exhaust gases is smaller than its weight, then the rocket will decelerate.

In space, when the rocket engines exert a backward force on the exhaust gases, then the exhaust gases exert a forward force on the rocket engines, causing forward acceleration. The presence of air or gases surrounding the rocket is not required. This allows spacecraft and satellites to be maneuvered in space. Small rocket engines (known as thrusters) are attached and used to allow the to spacecraft's position or orientation to be adjusted.

EXAMPLE 1

The total mass of the Atlas V rocket which carried the Mars Science Laboratory into space was $5\cdot31 \times 10^5$ kg.

At lift-off, the total thrust on the rocket provided by its rocket motors was $1\cdot02 \times 10^7$ N.

(a) Calculate the weight of the rocket.

$W = mg = 5\cdot31 \times 10^5 \times 9\cdot8 = 5\cdot20 \times 10^6$ N

(b) Draw a sketch of the rocket and label the forces acting on it at lift-off.

(c) Calculate the acceleration of the rocket at lift-off.

The unbalanced force F which acts on the rocket

$= $ thrust $-$ weight $= 1\cdot02 \times 10^7 - 5\cdot20 \times 10^6$

$\qquad\qquad\qquad = 5\cdot0 \times 10^6$ N upwards

$a = \dfrac{F}{m} = \dfrac{5\cdot0 \times 10^6}{5\cdot31 \times 10^5} = 9\cdot42$ m s^{-2} (upwards)

(d) Discuss why the value of the rocket's acceleration at lift-off calculated in part (c) is likely to change as the rocket ascends.

As the rocket ascends:

- the rocket's mass decreases as its fuel is used up
- the value for the gravitational field strength decreases.

These factors mean that the weight of the rocket reduces. This results in an increase in the unbalanced upward force, causing the acceleration to increase.

Rocket landing

When descending to land on a planet's surface, there are two forces acting on the rocket: the weight, and the thrust (upward force from the exhaust gases).

thrust $1\cdot02 \times 10^7$ N

weight $5\cdot20 \times 10^6$ N

contd

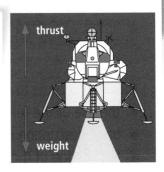

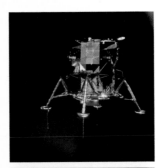

EXAMPLE 2

The total mass of the Apollo 11 lunar module which descended and landed on the surface of the Moon on 21 July 1969 was 1.51×10^4 kg. During the descent, close to the Moon's surface, a rocket engine producing a thrust of 4.5×10^4 N slowed the module before landing.

(a) Calculate the weight of the module on the Moon.

The gravitational field strength for the Moon is given on the Data sheet.

$W = mg = 1.51 \times 10^4 \times 1.6 = 2.42 \times 10^4$ N

(b) Label the diagram to show the forces and their direction acting on the module as it descends.

(c) Calculate the deceleration of the module.

The unbalanced force F which acts on the rocket
= thrust − weight = $4.5 \times 10^4 - 2.42 \times 10^4$

$F = 2.08 \times 10^4$ N (upwards)

$a = \dfrac{F}{m} = \dfrac{2.08 \times 10^4}{1.51 \times 10^4} = 1.38$ m s^{-2} (upwards)

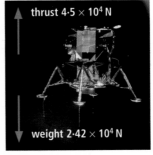

thrust 4.5×10^4 N

weight 2.42×10^4 N

The thrust is greater than the weight – this results in an unbalanced upward force which causes the rocket to decelerate as it approaches the planet surface

VIDEO LINK

Watch a video of the landing of the Apolo 11 lunar module at www. brightredbooks.net/ N5Physics

Gravitational field strength in different locations in the universe

The force of gravity is zero in deep space at distances well away from the Earth and other planets and stars. Different places in the solar system have different gravitational field strengths (g). This table appears in the Data sheet.

EXAMPLE 3

The weight on the surface of Mars of the Curiosity Rover robotic explorer vehicle is 3.33×10^3 N. Calculate the mass of the vehicle.

The gravitational field strength of Mars is obtained from the Data sheet.

$m = \dfrac{W}{g} = \dfrac{3.33 \times 10^3}{3.7} = 900$ kg

Gravitational field strength on the surface in Nkg-1	
Earth	9.8
Jupiter	23
Mars	3.7
Mercury	3.7
Moon	1.6
Neptune	11
Saturn	9.0
Sun	270
Uranus	8.7
Venus	8.9

EXAMPLE 4

On which planet would a falling ball accelerate at 8.9 ms^{-2}?

Venus. (The planet Venus has gravitational field strength of 8.9Nkg^{-1})

EXAMPLE 5

An astronaut's mass on Eart is 65 kg.

(a) What is the weight of the astronaut: (i) on Earth (ii) on the Moon?

   (i) $W = mg = 65 \times 9.8 = 637$ N; (ii) $W = mg = 65 \times 1.6 = 104$ N

(b) What is the mass of the astronaut on the Moon?

   Mass on Moon is unchanged at 65 kg.

EXAMPLE 6

In 2014, a spacecraft called the Rosetta successfully landed a probe of mass 100 kg on Comet 67P. The gravitational field strength of the comet was estimated at 5.4×10^{-5} Nkg^{-1}. Calculate the weight of the probe on the surface of the planet.

   $W = mg = 100 \times 5.24 \times 10^{-5} = 5.24 \times 10^{-3}$ N

Curiosity Rover on Mars.

THINGS TO DO AND THINK ABOUT

Research NASA's 'Journey to Mars' mission to send humans to Mars in the 2030s. In paricular, investigate the 'Space Launch System' – an advanced launch vehicle.

COSMOLOGY 1

The key concepts to learn in this topic are:
- the use of the term 'light year' and conversion between light years and metres
- basic description of the 'Big Bang' theory of the origin of the universe
- knowledge of the approximate estimated age of the universe.

The Andromeda galaxy is approximately 2·5 million light years from Earth

LIGHT YEAR

The light year is a unit of distance used in astronomy. A light year is defined as the distance that light travels in one year.

$d = vt$ so 1 light year $= 3 \times 10^8 \times 365{\cdot}25 \times 24 \times 60 \times 60$
$$= 9{\cdot}47 \times 10^{15}\,\text{m}$$

(Note that for this calculation 365·25 days – the average number of days in a year – is used.)

To reach the Earth, light takes approximately:
- 8 minutes from the Sun
- 4·3 years from the next nearest star
- 100 000 years from the edge of our galaxy.

EXAMPLE 1

Scientists believe that a black hole exists inside the Orion nebula, at a distance of 7800 light years from Earth.
(a) Change this distance into metres.

$d = vt = 3 \times 10^8 \times 7800 \times 365{\cdot}25 \times 24 \times 60 \times 60 = 7{\cdot}4 \times 10^{19}\,\text{m}$

(b) Explain why scientists use light years instead of metres when describing distances between stars and galaxies.

When huge distances, such as those connected with stars and galaxies, are calculated in metres, it is difficult to make comparisons because all of the results are so large. Light years are easier to interpret and compare.

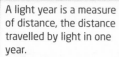

DON'T FORGET

A light year is a measure of distance, the distance travelled by light in one year.

A representation of the big bang

THE 'BIG BANG' THEORY

Origins of the observable universe

The current theory of the formation of the universe is sometimes described as the big bang model. This is a model or explanation of how the universe, which includes all matter and energy, began.

The explanation proposes that about 13·7 billion years ago there was an 'event', sometimes called a *singularity*. All of the matter and energy which now exists in our universe was contained in a single, extremely hot and dense point.

At this time, it is *not* thought that solid objects, including stars and galaxies, were produced. Instead, the model suggests that there were distinct stages in time after the big bang when certain events occurred. This is a simplified version of the model:

- After the singularity, it is suggested that the universe started expanding very rapidly in all directions; this period of expansion is known as inflation and it was powered by massive energy, known as primordial energy. The form of this energy is unidentified in the current model.
- Scientists believe that during this inflation period, this energy was distributed evenly throughout the universe. (Data received from a study of the **Cosmic Microwave**

contd

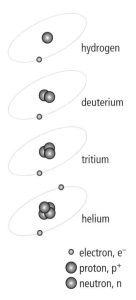

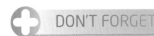

The first atoms to be produced, according to the big bang model

Background Radiation (CMBR) in space, shows that the universe today has an even distribution of energy throughout, with some fluctuations). This evidence supports the Big Bang theory.

- The universe continued to expand and cool, eventually protons and neutrons formed.
- Around 400 000 years after the big bang, further cooling and expanding led to the formation of atoms. Hydrogen and its isotopes, deuterium and tritium, and also helium atoms were produced.
- Following the formation of these atoms, expansion and cooling continued. This stage in the development of the universe is known as the Dark Ages because there were no single objects which produced any light (or other EM radiation).
- The first stars were formed when the hydrogen and helium atoms collected together under gravitational forces, until each mass was so great that nuclear fusion started in the core. Fusion reactions released the energy to 'ignite' the stars. Massive stars were produced with short lives. These exploded as supernovae and caused huge ionisation.
- It is thought that the supernovae collapsed to form black holes which increased in size to become the huge black holes found at the centre of some galaxies today.
- The eventual formation of stars and galaxies followed, which continue to move outwards as part of the big bang expansion process.

Evidence for the big bang theory

Evidence from space exploration supports much of the big bang model. However, there are many parts of the model which still require further investigation and data to fully explain. There is evidence to show that the universe is still expanding.

However, an expanding universe requires energy to sustain the expansion. This is because the force of gravitational attraction between galaxies opposes this 'outward expansion'. This energy which keeps the universe expanding has yet to be identified, and has been named **'dark energy'** since its source has, so far, to be discovered. **Observations** of **stars** in **orbit** within their **galaxies** have indicated that the observed mass of the galaxies are too small to support the observed orbiting velocities.

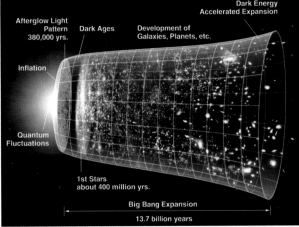

This unaccounted for mass has been named **'dark matter'**, since its nature has, so far, not been discovered. No direct measurements of dark matter have so far been made.

Experiments carried out at at the Large Hadron Collider (LHC) may provide more direct clues about dark matter.

Current measurements estimate that the universe is comprised of less than **5% conventional matter** and **27% of dark matter** and **68% dark energy**.

There is evidence to show that the universe is still expanding

This is a simplified explanation of the theory of how the universe began. When the **Big Bang model** was proposed, there was limited evidence to support it.

Now, a lot of research and observation is being carried out into the existing universe to verify, and sometimes change, more of the complex detail of the Big Bang model.

 THINGS TO DO AND THINK ABOUT

The timescale for the theory of the big bang estimates when certain 'key' events took place. Carry out research to find out when the formation of the Milky Way galaxy took place.

COSMOLOGY 2

The key concept to learn in this topic is:

- awareness of the use of the whole electromagnetic spectrum in obtaining information about astronomical objects.

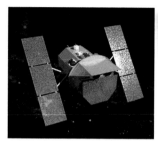

The Swift telescope detects gamma ray bursts from black holes

USING THE WHOLE ELECTROMAGNETIC SPECTRUM TO OBTAIN INFORMATION ABOUT ASTRONOMICAL OBJECTS

Radiation received from astronomical objects in space covers the entire electromagnetic spectrum range. This radiation travels at the speed of light, but even travelling at this speed, the radiation takes a long time to travel the huge distances across the universe.

The radiation detected provides information about what was happening a long time ago in the universe. This information can be analysed; it is used to support our understanding of the universe and how it was formed.

A range of different detectors is required to detect radiation signals of different frequencies.

Telescopes designed to detect radiation from specific bands of the EM spectrum are used to obtain information from from objects in space.

Earth-based detectors are used for radiations which can penetrate the Earth's atmosphere

Satellites are used to carry detectors which can receive signals that cannot reach the Earth's surface or that are too small to detect on Earth

ONLINE

Learn more by reading NASA's 'Gamma Ray Telescopes and Detectors' at www.brightredbooks.net/N5Physics

TYPES OF TELESCOPE

Gamma ray telescopes

The Earth's atmosphere absorbs much of the gamma radiation from space. So, the gamma radiation from space is studied using gamma ray telescopes carried by satellites in orbit *above* the atmosphere, and also by some ground based detectors.

The Fermi satellite uses a gamma ray telescope to investigate sources of **cosmic rays**. Cosmic rays are thought to be important in the production of gamma radiation from different astronomical objects, such as **supernovae** remains and **black holes**. Investigating the sources of cosmic rays provides data which improves our understanding of the structure and mechanism of supernovae events, and supports the 'Big Bang' theory.

X-ray telescope

X-ray telescopes

The Earth's atmosphere absorbs many of the X-rays from space. So, these are studied using X-ray telescopes that are carried by satellites in orbit above the atmosphere. Chandra, XMM-Newton and Suzaku are satellite observatories which have X-ray telescopes on board.

contd

Data received from outside our galaxy using these X-ray telescopes indicates the presence of a previously undetected massive cloud of very hot gas. The existence of this cloud is regarded as an important piece of evidence supporting the **big bang model** which describes the origin of the universe.

Ultraviolet telescopes

Data from **ultraviolet radiation** detected from space by the GALEX satellite has contributed to research into how stars are formed inside galaxies. The Hubble satellite also contains an ultraviolet telescope.

Optical telescopes

Light from stars and galaxies viewed through optical telescopes can be analysed to obtain line spectra which identify the chemical elements present in the stars.

Observation and analysis of the motion of distant galaxies suggests that most galaxies are moving *away* from Earth. This supports the 'expanding universe theory' of the standard big bang model.

Infrared telescopes

Study of space using optical telescopes is limited because some astronomical objects lie behind dense regions of dust and gas. Infrared radiation has longer wavelengths than visible light which allow it to travel through these regions of space without being absorbed or scattered.

Infrared telescopes have been used to investigate the structure of distant stars and galaxies. The Spitzer satellite has an infrared telescope which has obtained data that has increased astronomers' understanding of star formation within galaxies.

Radio telescopes

Radio telescopes used to detect radio waves from space can be land-based, or carried on satellites.

Radio waves are emitted from objects in space. Radio telescopes have been used to detect and receive radio waves from pulsars and quasars. They have allowed production of maps showing the positions of galaxies and nebulae.

COBE and WMAP are satellites which carry radio telescopes. They have obtained data which has provided astronomers with more information about the age of the universe, and the origin and structure of galaxies.

These telescopes gather data from the visible part of the EM spectrum

Infrared telescopes used to detect infrared radiation from space can be land based like this one, or carried on satellites

The Spitzer satellite detects infrared radiation from deep space

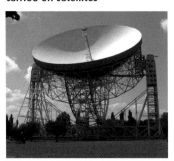

A land-based radio telescope

 DON'T FORGET

Using satellites to explore and gather data from space has increased our understanding of planet Earth.

 ONLINE

Check out 'Infrared Telescopes Spy Small, Dark Asteroids' at www.brightredbooks.net/N5Physics

DON'T FORGET

Some satellites are used to detect electromagnetic radiation which cannot penetrate the Earth's atmosphere.

 ONLINE TEST

How well have you learned about the different types of telescopes? Take the 'Space Exploration' test at www.brightredbooks.net/N5Physics

 ## THINGS TO DO AND THINK ABOUT

1. There are many different areas of space research being carried out to improve our current understanding of the universe, especially the standard big bang model. Carry out your own research into one aspect of the big bang model, considering the value of this research. Include a discussion of whether it adds to the understanding of the universe, or contradicts previously understood ideas.

2. Sub-atomic particles feature in the theory of the standard model for the big bang. Find out what these particles are and how they are being investigated today, for example, in the Large Hadron Collider.

3. Investigate the claim by astronomer Edwin Hubble that the velocity of a galaxy depends on its distance from Earth.

COSMOLOGY 3

The key concepts to learn in this topic are:
- the identification of continuous and line spectra
- the use of spectral data for known elements, to identify the elements present in stars.

CONTINUOUS AND LINE SPECTRA

As we have seen, when white light passes through a three dimensional triangular shaped piece of glass known as a *prism*, it separates into different colours. These colours correspond to the different frequencies of visible light.

Used in this way, the glass prism functions as a simple *spectroscope* – it enables us to identify the different frequencies of light present in a source. The prism *refracts* each colour by different amounts: violet light (highest frequency) is the colour that is refracted most and red light (lowest frequency) is refracted least.

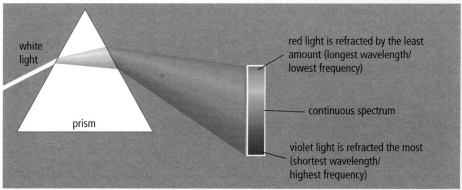

white light

prism

red light is refracted by the least amount (longest wavelength/ lowest frequency)

continuous spectrum

violet light is refracted the most (shortest wavelength/ highest frequency)

A prism separates the component colours of white light

Light from stars, galaxies and nebulae can be studied using a spectroscope containing a prism. This separates the light from, for example, the star into its various constituent colours (or frequencies).

Line emission spectra

When heated, each element or compound emits a characteristic spectrum composed of light of particular frequencies. This is called a *line emission spectrum*. The elements and compounds that make up a star (or a light-emitting exoplanet) can be identified by the lines emitted. The diagram below shows the line emission spectra for the elements sodium, mercury and neon.

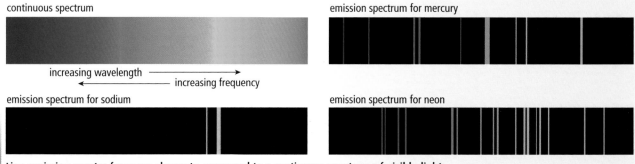

continuous spectrum

increasing wavelength

increasing frequency

emission spectrum for mercury

emission spectrum for sodium

emission spectrum for neon

Line emission spectra for some elements, compared to a continuous spectrum of visible light

DON'T FORGET

The line spectra from stars can be used to identify the elements present.

Line spectra can be obtained for other parts of the electromagnetic spectrum, as well as for visible light. Infrared light from exoplanet HR 8799c, discovered in orbit around the star HR 8799, was obtained using the Keck II infrared telescope in Hawaii. Analysis of

contd

this infrared light produced line spectra that identified the presence of carbon monoxide and water on the planet. Such information can be used to determine how the planet was formed, and could even reveal the correct conditions for the possible presence of life on distant planets.

EXAMPLE:

Some spectral lines from a distant star are shown. Identify the elements present in the star.

The elements are hydrogen and helium. (Hint: use a ruler to 'line up' the spectral lines of the elements with the star spectrum to find which elements match.)

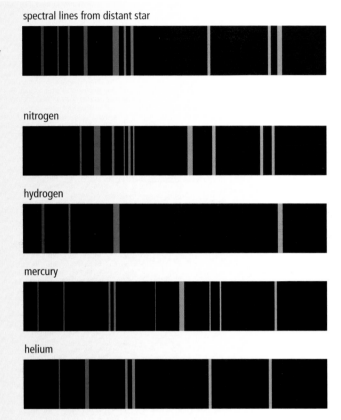

spectral lines from distant star

nitrogen

hydrogen

mercury

helium

Line absorption spectra

A line absorption spectrum is obtained when white light from stars passes through gas clouds (nebulae) in space.

When gas clouds absorb specific frequencies of light, *black lines* appear in the *continuous spectrum* of emitted light. This absorption spectrum can be used to identify the elements present in the cloud.

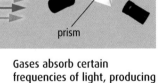

Gases absorb certain frequencies of light, producing an absorption spectrum

Frequencies in the red–yellow, green and indigo regions of the spectrum have been absorbed, as indicated by the black lines

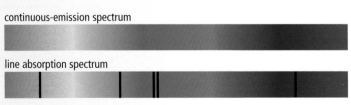

continuous-emission spectrum

line absorption spectrum

THINGS TO DO AND THINK ABOUT

1. Investigate current research into the structure of the universe. Find out if improvements or changes to the 'big bang model' have been discovered. (Try trusted websites like NASA and BBC/science.)

2. Research which scientists first discovered the line spectra of elements. Which element was first discovered in stars through spectrometry before being discovered on Earth?

ONLINE TEST

How well have you learned about continuous and line spectra? Take the 'Cosmology' test at www.brightredbooks.net/N5Physics

ELECTRICITY

ELECTRICAL CHARGE CARRIERS

The key concepts to learn in this topic are:

- definition of electrical current as the electric charge transferred per unit time
- use of an appropriate relationship to solve problems involving charge, current and time: Q = It
- knowledge of the difference between alternating and direct current
- identification of a source (as a.c. or d.c.) based on oscilloscope trace or image from data logging software.

ELECTRIC CHARGE

Electrons are negatively charged and protons are positively charged. The diagram shows that, when a plastic rod is rubbed with a duster, the duster gains electrons from the **atoms** of the plastic rod. The plastic rod becomes positively charged because it has lost negatively charged electrons.

The symbol for charge is Q and charge is measured in coulombs (C). When charged particles with the same charge are close to each other, there is a repelling **force** between them. Charged particles with opposite charge attract each other.

This attraction of unlike charges is used in lots of modern devices, such as photocopiers, laser printers and paint sprayers.

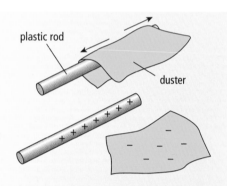

plastic rod
duster

Transfer of negative charge from rod to duster

Like charges repel!

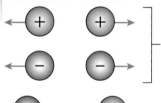

repelling forces

attracting forces

Forces between like and unlike charges

ELECTRIC CURRENT

When electrons move through a conductor, for example a copper wire, an electric current is produced.

electrons

Electrons in a wire

Current (I) is a measure of the amount of charge (Q) moving, or transferred, through a conductor every second (t). When a charge of 1 coulomb (1 C) is transferred through a conductor each second, there is a current of 1 ampere (1 A). Charge is calculated using the relationship:

$$\text{charge} = \text{current} \times \text{time} \qquad Q = It$$

EXAMPLE 1

Calculate how much charge passes through the wire in 4 minutes.

$Q = It$

Remember to convert t into seconds by multiplying by 60.

$Q = 5 \times 4 \times 60 = 1200\,\text{C}$

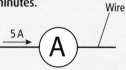

Wire

5 A (A)

contd

EXAMPLE 2

The current in the wire connected to an electric kettle is 5·5 A.
Calculate how long it takes a charge of 2200 C to pass a point in the wire.

$$Q = It, \; 2200 = 5{\cdot}5t \,, t = \frac{2200}{5{\cdot}5} = 400 \; s$$

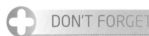

DON'T FORGET

Direct current is in one direction only; alternating current changes direction.

ALTERNATING AND DIRECT CURRENT

When charge moves constantly in the same direction through a conductor, this is called direct current (d.c.); when charge moves back and forwards through the conductor, this is called alternating current (a.c.).

A battery is a d.c. electrical source. The mains supply of electricity in Britain is a.c. The declared value of the mains voltage is 230 V, the frequency of the alternating current is 50 Hz.

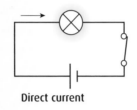

Direct current

live

neutral **Alternating current**

Identifying d.c and a.c sources

An electrical power source can be investigated using an oscilloscope or data logging software connected to a sensor.

Controls on the oscilloscope can be adjusted to allow the amplitude (voltage) and the frequency of an electrical a.c. signal to be measured.

The Y axis indicates the voltage amplitude.

The X axis indicates the time for the alternating signals to be produced.

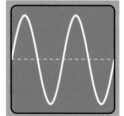

d.c. display

A d.c. electrical signal displayed on an oscilloscope. The signal is a horizontal line above (or below) the central axis.

An a.c. electrical signal displayed on an oscilloscope. The signal is a sine wave.

a.c. display

EXAMPLE:

The electrical signal of an a.c. supply is displayed on the oscilloscope grid. The grid is made from 1 cm squares. The voltage and time scale settings are shown.

(a) Determine the amplitude of the voltage supply.

Y axis setting: 1·5 V per cm

so $V_{amplitude}$ = 1·5 × 4 boxes = 6V

(b) Determine the frequency of the supply.

X axis setting: 0·6 ms per cm

so time for one wave,

T = 0·6 × 6 boxes = 3·6 ms

$$T = \frac{1}{f} \rightarrow 3{\cdot}6 \times 10^{-3} = \frac{1}{f}$$

$$f = \frac{1}{3{\cdot}6 \times 10^{-3}} = 278 \; Hz$$

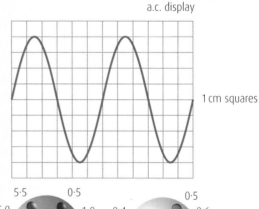

1 cm squares

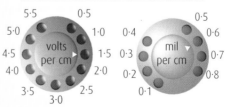

Y (vertical) axis   X (horizontal) axis
(volts per centimetre) (milliseconds per centimetre)

THINGS TO DO AND THINK ABOUT

Many domestic appliances operate using a.c. power supplies. Make a list and investigate the electrical and electronic devices which you personally use. Safely find out which are operated on a.c. and which are d.c.

POTENTIAL DIFFERENCE (VOLTAGE)

The key concepts to learn in this topic are:

- knowledge that a charged particle experiences a force in an electric field
- knowledge of the path a charged particle follows: between two oppositely charged parallel plates; near a single point charge; between two oppositely charged points; between two like charged points
- knowledge that the potential difference (voltage) of the supply is a measure of the energy given to the charge carriers in a circuit.

VIDEO LINK

To see the effect of an electric field, watch this video at www.brightredbooks.net/N5Physics

THE PATH OF CHARGED PARTICLES IN ELECTRIC FIELDS

An **electric field** is an area where charged particles (such as protons and electrons) experience a force. An electric field can be produced by applying a voltage to two metal plates that are separated by a gap, as shown in the diagram.

If charged particles are present in the gap, they experience a force and move.

Electric fields can be represented by drawing lines between the plates in the direction of positive charge to negative charge. These electric field lines indicate the direction of travel that a **positively** charged particle would have if placed inside the electric field. Negatively charged particles placed inside the electric field would travel in the opposite direction.

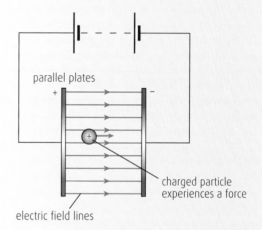

parallel plates

charged particle experiences a force

electric field lines

DON'T FORGET

Charged particles, such as electrons, protons, **alpha particles**, charged paint droplets and smoke particles, all experience a force in an electric field.

The electric field lines are shown:

- for single positively and negatively charged points;
- between two oppositely charged points;
- between two like charged points

The path taken by a positively charged particle near these electric fields is also shown.

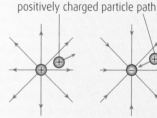

positively charged particle path

Electric field lines for a single positively charged point

Electric field lines for a single negatively charged point

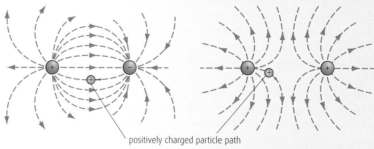

positively charged particle path

electric field lines for oppositely charged points

electrical field lines for like charged points

contd

EXAMPLE

The atomic particles neutrons, protons and electrons are directed into an electric field between two metal plates as shown.

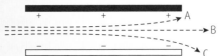

Explain which path is taken by each particle.

A   Electron, the negatively charged particle experiences a force upwards and moves towards the positive plate.

B   Neutron, the neutron has no charge, does not experience any force and so follows its original path.

C   Proton, the positively charged particle experiences a force downwards and moves towards the negative plate.

POTENTIAL DIFFERENCE (VOLTAGE)

When a **battery** is connected to a wire, this produces an electric field in the wire.

The wire in the diagram is made from metal. The atoms in metals have electrons which are free to move from atom to atom. These electrons are charged carriers which experience a force causing them to move through the wire. Each time the electrons move through the supply battery, it transfers **electrical energy** to them.

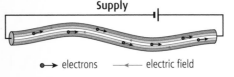

Charged particles moving through a wire

Electrons which have received electrical energy from the battery move through wires around the circuit. Work is done each time these charge carriers pass through a component in the circuit. Some of their energy is transferred to the component. This causes an energy difference across the component, which is known as an electrical **potential difference (p.d.)**

The energy which is required to move the electrons through the component is a measure of the potential difference across the component; it is measured in volts. When 1 joule (1 J) of energy is required to move 1 coulomb (1 C) of charge through a component, the potential difference is 1 joule per coulomb ($1\,JC^{-1}$), or 1 volt (1 V).

The potential difference (voltage) of the supply is a measure of the energy given to the charge carriers (electrons) in a circuit.

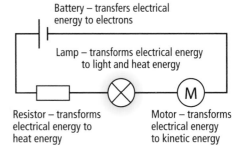

Battery – transfers electrical energy to electrons

Lamp – transforms electrical energy to light and heat energy

Resistor – transforms electrical energy to heat energy

Motor – transforms electrical energy to kinetic energy

Transformation of energy within the components of a circuit

THINGS TO DO AND THINK ABOUT

1. Research the variety of uses of electric fields including how electric fields are responsible for accelerating protons to 99·9% of the speed of light in the Large Hadron Collider and how electric fields are used in LCD televisions.

OHM'S LAW 1

The key concepts to learn in this topic are:
- calculation of the gradient of the line of best fit on a V-I graph to determine resistance
- use of appropriate relationships to solve problems involving potential difference (voltage), current and resistance:

$$V = IR, \quad V_2 = \left(\frac{R_2}{R_1 + R_2}\right)V_s, \quad \frac{V_1}{V_2} = \frac{R_1}{R_2}$$

- knowledge of the qualitative relationship between the temperature and resistance of a conductor
- description of an experiment to verify Ohm's law.

AN OVERVIEW OF OHM'S LAW

Resistance is the property of a conductor to oppose current. The larger the resistance in a circuit, the smaller the current. Resistance can be calculated using the relationship:

$$R = \frac{V}{I}$$

Where: R is the resistance in ohms (Ω)
V is the potential differences (voltage) in volts (V)
I is the current in amperes (A)

This is Ohm's Law.

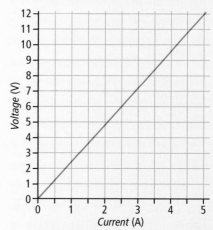

Resistors

DON'T FORGET

In physics, a conductor is the general name given to any component which allows current through it, for example, a wire, resistor or lamp.

An experiment to verify Ohm's Law

An experiment using the apparatus in Circuit 1 can be used to verify Ohm's Law for a conductor (a conductor means any electrical component which conducts, including resistors).

- Adjust the variable resistor to change the current in the circuit and the voltage across the conductor.
- Record the ammeter and voltmeter readings for each adjustment in a table, like the one shown to the left.
- Draw a V-I graph of the results.
- A straight line graph which passes through the origin should be obtained. This shows that the current is directly proportional to the voltage.

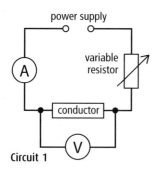

Circuit 1

Voltage (V)	Current (A)	$\frac{V}{I}$
3·0	1·25	2·4
6·0	2·5	2·4
9·0	3·75	2·4
12·0	5·0	2·4

Calculation of the gradient of the line of the best fit on a V-I graph to determine resistance

A graph can be drawn using the results of an experiment to verify Ohm's Law. In the V-I table shown, a third column gives the results of a calculation of $\frac{V}{I}$ for each pair of V-I readings.

The V-I graph drawn here is the line of the best fit using these pairs of readings.

A constant value for $\frac{V}{I}$ is obtained for each pair of readings. This value is known as the resistance, R, of the conductor.

This straight line graph which passes through the origin shows that the current is directly proportional to the voltage (the graph shows a linear relationship). This means that if voltage is doubled, current is doubled, and so on.

VIDEO LINK

Check out the video about Ohm's Law at www.brightredbooks.net/N5Physics

A graph of V–I can be used to prove Ohm's Law

contd

The *gradient* of this graph will determine the ratio $\frac{V}{I}$.

To calculate the gradient of the *V–I* graph, use :

$m = \dfrac{y_2 - y_1}{x_2 - x_1}$ where m is the gradient.

Choose values for (x_2, y_2) and (x_1, y_1) from the graph. (Hint: choose pairs of values which are easy to read from the graph!)

Choosing: $(x_2, y_2) \rightarrow (5\cdot0, 12\cdot0)$ and $(x_1, y_1) \rightarrow (1\cdot5, 3\cdot6)$

$m = \dfrac{y_2 - y_1}{x_2 - x_1} = \dfrac{12\cdot0 - 3\cdot6}{5\cdot0 - 1\cdot5} = \dfrac{8\cdot4}{3\cdot5} = 2\cdot4$

The gradient is the resistance, *R*, of the conductor, so $R = 2\cdot4\,\Omega$.

The results and graph obtained from this experiment confirm the relationship between voltage *V*, current *I* and resistance *R*: $R = \dfrac{V}{I}$

ONLINE TEST

For a test on Ohm's Law, visit www.brightredbooks.net/N5Physics

EXAMPLE 1

The following V-I graph was drawn using the results of an experiment to determine the resistance of a resistor.

Use the graph to determine the resistance.

resistance = gradient of V–I graph

resistance $= \dfrac{V_1 - V_2}{I_1 - I_2} = \dfrac{5\cdot0 - 1\cdot8}{0\cdot28 - 0\cdot1} = 17\cdot8\,\Omega$

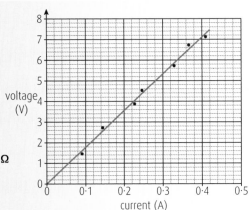

EXAMPLE 2

The graph shows how the current is related to the applied voltage across two separate resistors, X and Y.

Which row shows the correct values of the two resistors X and Y?

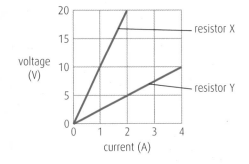

	Resistance of X (Ω)	Resistance of Y (Ω)
A	0.1	0.4
B	10	2.5
C	5	5
D	20	10
E	10	20

Select values for V and I from the graphs to calculate the resistance of X and Y:

for X: $R = \dfrac{V_1 - V_2}{I_1 - I_2} = \dfrac{20 - 0}{2 - 0} = 10\ \Omega$

for Y: $R = \dfrac{V_1 - V_2}{I_1 - I_2} = \dfrac{10 - 0}{4 - 0} = 2\cdot5\ \Omega$

so answer B is correct.

 THINGS TO DO AND THINK ABOUT

When using the apparatus described on page 50 to determine the value, R, of a resistor, it is more accurate to obtain several values of V and I; and then to calculate the gradient of a V-I graph to find R. Find out and explain why this is more accurate for this experiment.

OHM'S LAW 2

The key concept to learn in this topic is:

- use of appropriate relationships to solve poblems involving potential difference (voltage), current and resistance.

1. USING OHM'S LAW V = IR

Calculate the current in the circuit shown.

$$I = \frac{V}{R} = \frac{24}{3.3 \times 10^3} = 7.3 \times 10^{-3} \text{ A}$$

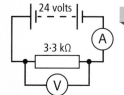

24 volts

3·3 kΩ

Calculate the supply voltage in the circuit shown.

$$V = IR = 2.8 \times 12.5 = 35 \text{ V}$$

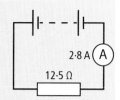

2·8 A

12·5 Ω

Calculate the value of resistor R in the circuit shown.

$$R = \frac{V}{I} = \frac{6.0}{2.5 \times 10^{-3}} = 2.4 \times 10^3 \, \Omega$$

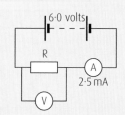

6·0 volts

R

2·5 mA

2. USING $V_2 = \left(\frac{R_2}{R_1 + R_2}\right) V_s$ and $\frac{V_1}{V_2} = \frac{R_1}{R_2}$ IN VOLTAGE DIVIDER CIRCUITS

Whenever components are used in circuits and are connected to a battery, they can be connected in series with one or more resistors, in a circuit known as a *voltage divider circuit*.

The battery voltage divides between the resistor and the component. This is to control the voltage across the component – sometimes to protect it from damage and sometimes to be able to monitor how the voltage across the component changes.

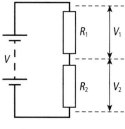

Examples of voltage divider circuits

Voltage divider circuit   LED circuit   LDR circuit   Thermistor circuit

A simple voltage divider circuit consists of two resistors in series connected to a supply. The resistor with the greatest resistance has the greatest share of the supply voltage across it. The voltages across the components add up to the supply voltage.

When calculating values of voltage or resistance in these circuits, there are *two* relationships which can be used, depending on the information given in the circuit.

Voltage divider circuit

Method 1

When the supply voltage and both resistor values are given, and either V_1 or V_2 has to be calculated, this relationship is used:

$$V_2 = \left(\frac{R_2}{R_1 + R_2}\right) V_s$$

Calculate V_2.

$$V_2 = \left(\frac{R_2}{R_1 + R_2}\right) V_s$$

$R_1 = 229 \, \Omega$,

$R_2 = 330 \, \Omega$

$$V_2 = \left(\frac{330}{229 + 330}\right) \times 4.5$$

$$= 2.66 \text{V}$$

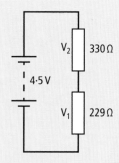

V_2 330 Ω

4·5 V

V_1 229 Ω

contd

Method 2

This relationship is used when the supply voltage is not given, but the voltage across a resistor or a resistor value has to be calculated:

$$\frac{V_1}{V_2} = \frac{R_1}{R_2}$$

Calculate the value of R_2.

$$\frac{V_1}{V_2} = \frac{R_1}{R_2} \quad V_1 = 8\cdot9\text{V}, \ V_2 = 3\cdot2\text{V}$$

$$\frac{8\cdot9}{3\cdot2} = \frac{420}{R_2}$$

$$R_2 = 420 \times \frac{3\cdot2}{8\cdot9}$$

$$= 151\,\Omega$$

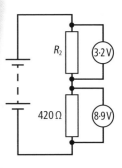

THE RELATIONSHIP BETWEEN RESISTANCE AND TEMPERATURE CHANGE

The resistance of a conductor remains constant as long as its temperature remains constant. Materials which have a linear relationship between the applied voltage and current are said to be *ohmic conductors*. Conductors made from carbon or metals are ohmic over a range of current values.

NON-OHMIC CONDUCTORS

For some materials, a change in temperature can cause a change in resistance. Materials which have a non-linear relationship between the applied voltage and current are said to be *non-ohmic conductors*. The shape of the *V–I* graph for non-ohmic conductors is not a straight line. For non-ohmic conductors, the value of resistance cannot be found by calculating the gradient of a voltage–current graph. It has to be calculated using Ohm's Law with specific voltage and current values taken from the graph or table of results. Non-ohmic conductors do not have a fixed resistance when the current in the conductor changes.

Examples of non-ohmic components

When a **filament lamp** is investigated using the circuit shown in the diagram this voltage–current graph is obtained.

The graph is not a straight line, which means that the resistance of the filament wire is not constant for different current values. As the current increases, the filament heats up, and the resistance *increases*.

When a **thermistor** is investigated using the circuit in the diagram this voltage–current graph is obtained.

The graph is not a straight line, which means that the resistance of the thermistor is not constant for different values of current. As the current increases the resistance of the thermistor *decreases*.

Use the following *V–I* graph of a thermistor to calculate its resistance when the current is 2 A, 5 A and 8 A.

$$2\,\text{A}: R = \frac{V}{I} = \frac{10}{2} = 5\,\Omega \quad 5\,\text{A}: R = \frac{V}{I} = \frac{16}{5} = 3\cdot2\,\Omega$$

$$8\,\text{A}: R = \frac{V}{I} = \frac{21}{8} = 2\cdot6\,\Omega$$

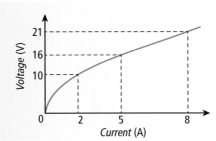

THINGS TO DO AND THINK ABOUT

Investigate the properties of a **diode** and find out how its resistance changes when different voltages are applied to it.

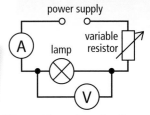

power supply

Filament lamp investigation

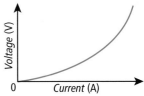

Voltage–current graph for filament lamp

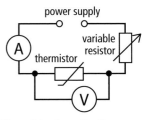

power supply

Thermistor investigation

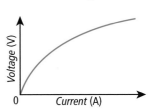

Voltage–current graph for thermistor

PRACTICAL ELECTRICAL AND ELECTRONIC CIRCUITS: MEASURING CURRENT, VOLTAGE AND RESISTANCE

The key concept to learn in this topic is:

- the measurement of current, potential difference (voltage) and resistance using appropriate meters in simple and complex circuits

DON'T FORGET

In series circuits, the voltages across each component add up to the supply voltage.

MEASURING CURRENT AND VOLTAGE IN SERIES CIRCUITS

In a series circuit, the *same current* passes through each component.

In a series circuit, the supply *voltage divides* across each component.

Current and voltage in components in series

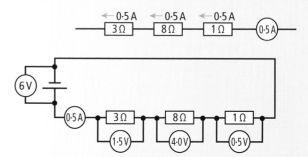

VIDEO LINK

To see the differences between series and parallel circuits in practice, watch the video at www.brightredbooks.net/N5Physics

MEASURING CURRENT AND VOLTAGE IN PARALLEL CIRCUITS

- In a parallel circuit, the supply *current splits up* through each component when it reaches a branch in the circuit.

- In a parallel circuit, the *voltage* across components connected in parallel *remains the same as the supply voltage.*

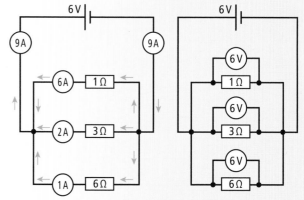

Current and voltage in components in parallel

In the circuits above, the 9 A supply current splits up through each branch of the circuit.

The supply voltage of 6 V appears across each resistor connected in parallel.

Note that for resistors connected in parallel, the smallest current is present in the branch with the largest resistance, and the largest current is present in the smallest resistance.

MEASURING CURRENT AND VOLTAGE IN COMPLEX CIRCUITS

A complex circuit contains some components that are connected in series and some that are connected in parallel. Care is required to determine the current and voltage at different positions in the circuit.

Current in a complex circuit

The circuit in the diagram consists of parts where resistors are connected in series, and parts where they are connected in parallel. To determine the current at different

contd

positions in the complex circuit, it is important to trace the path of current from the supply through the circuit and back to the supply.

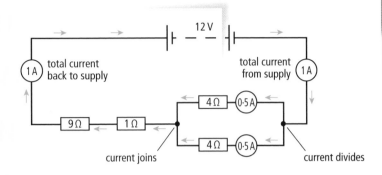

- In the circuit, the total current of 1 A leaves and enters the supply.

- The total current of 1 A is present in the 1 Ω and 9 Ω resistor because they are connected in series with the supply. This is shown on the **ammeters** which are displaying the total current.

- The 4 Ω resistors are connected in parallel. The total current of 1 A divides equally between them – 0·5 A and 0·5 A. This is displayed on the ammeters in the parallel part of the circuit.

Current in a complex circuit

Voltage in a complex circuit

To determine the voltage across components in a complex circuit, it is important to recognise the parts of the circuit where components are connected in series, and the parts that are connected in parallel.

In the circuit in the diagram the supply voltage will be distributed across the components.

Using voltmeters to display the voltage across different components shows that the voltage divides where there are components in series – across the 9 Ω and 1 Ω resistors. Components that are connected in parallel have the *same* voltage across both resistors.

Work out which parts of the circuit are in parallel and which parts are in series

DON'T FORGET

In complex circuits, current has the same value as the supply in the series parts, but divides in the parallel parts.

ONLINE TEST

Take the 'Practical Electrical and Electronic Circuits: Measuring Current, Voltage and Resistance' test at www.brightredbooks.net/N5Physics

MEASUREMENT OF RESISTANCE

There are two ways to measure the resistance of a component in a circuit.

Method 1 – remove component from the circuit and connect to an ohmmeter
For example, in the circuit shown above, if the 4 Ω resistance was unknown and had to be determined, we would disconnect the resistor and connect it to an ohmmeter as shown here.

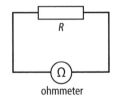

ohmmeter

Measured value for $R = 4\,\Omega$

Method 2 – use an ammeter and voltmeter, and a calculation
With the component connected in a circuit, an ammeter and voltmeter can be used to measure the current in it and the voltage across it. Then a calculation is carried out.

For example, we could connect the resistor to a supply, as shown in the diagram, and obtain current and voltmeter readings. Then we would use Ohm's Law to calculate the resistance.

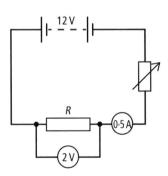

$$R = \frac{V}{I} = \frac{2}{0\cdot5} = 4\,\Omega$$

When using this method, it is good practice to repeat the resistance calculation for different readings of current and voltage, and to calculate the *average* value for R.

THINGS TO DO AND THINK ABOUT

1. When measuring current, voltage and resistance using a multimeter, it is important to adjust the multimeter for the appropriate measurement of I, V or R. Find out how to select the appropriate settings on a multimeter to measure current, voltage and resistance, and how to make sure that the correct *range* is selected for the measurement.

PRACTICAL ELECTRICAL AND ELECTRONIC CIRCUITS: STANDARD ELECTRICAL COMPONENTS 1

The key concepts to learn in this topic are:

- knowledge of the circuit symbol, function and application of standard electrical and electronic components: cell, battery, lamp, switch, resistor, voltmeter, ammeter, LED, motor, microphone, loudspeaker, photovoltaic cell, fuse, diode, capacitor, thermistor, LDR, relay, transistor
- for transistors, knowledge of the symbols for an npn transistor and an n-channel enhancement mode MOSFET. Explanation of their function as a switch in transistor switching circuits.

DON'T FORGET

You could be asked to draw these symbols in the exam.

CIRCUIT SYMBOLS

The common standard electrical and electronic components and symbols used in this course are shown in the table.

Component	Property	Function	Symbol
Cell	A store of *chemical energy* which transforms into electrical energy	Provides electrical energy to make charge move in a circuit	
Battery	A store of chemical energy which transforms into electrical energy	Provides electrical energy to make charge move in a circuit	
Photovoltaic cell (solar cell)	Transforms *light energy* into electrical energy	A renewable energy source, these cells can be used to provide electrical energy to recharge batteries for portable electronic appliances and, for example, vehicle speed warning signs	solar cell
Variable resistor	Opposes the movement of charge in a circuit	Used to vary the size of current	
Voltmeter	Measures voltage in circuits	Measures the voltage across components	
Ammeter	Measures current in circuits	Measures the current in components	
Fuse	Device which contains a wire with a low melting point which heats up and melts if current exceeds a given value	The fuse melts and breaks the circuit if the current in the circuit increases above the set fuse value	
Diode	Electronic device which allows current in one direction only	Used in electronic circuits	
LED (light emitting diode)	Electronic device which allows current in one direction only, emits light, and uses low values of current	Used as an indicator and for low energy lighting, e.g. car brakelights and sidelights	
Capacitor	Stores charge	Used in electronic amplifiers and timing circuits	
LDR (light dependent resistor)	Electronic device with changing resistance depending on the light level	Used in circuits to detect and control light levels	
Thermistor	Electronic device with changing resistance depending on its temperature	Used in circuits to detect and control temperature	
Motor	Transforms electrical energy into kinetic energy	Motor speed depends on size of the supply voltage and its direction can be reversed	
Loudspeaker	Transforms electrical energy signals into *sound energy*	Converts output from amplifiers into sounds	
Microphone	Transforms sound energy signals into electrical energy	Converts sound energy into electrical signals for amplifiers	
Relay	Acts as a remote switch to switch on a separate circuit	When current is present in the relay coil, a switch in a separate circuit is closed	coil
Transistor	Semiconductor device which acts as an electronic switch or amplifier	Used in electronic switching circuits and amplifiers	bipolar transistor MOSFET*

(*metal oxide semiconductor field-effect transistor)

TRANSISTORS

Transistors are electronic components that feature in many electrical devices including computers, televisions and mobile phones. Sometimes they are discrete components, like the ones shown in the picture, but often hundreds are contained in a single microchip.

- A transistor is a semiconductor device which can be used as a switch.

- Two of the main types of transistor are the bipolar transistor and the MOSFET.

When used in a switch circuit, the transistor is off (non-conducting) until its input voltage is greater than 0·7 V (for a bipolar transistor) or 2 V (for a MOSFET transistor), at which point the transistor switches on (starts conducting).

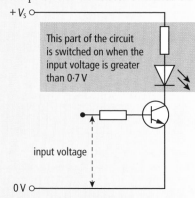

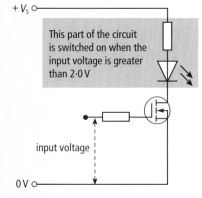

Bipolar transistor switching circuit　　**MOSFET transistor switching circuit**

Transistors are used to switch different devices on (or off) when their input voltage changes above (or below) the switching voltage. This changing voltage is often provided by a voltage divider circuit. The change in voltage can be caused by a variety of different conditions, such as a change in light level or temperature.

Bipolar transistor　　**MOSFET transistor**

VIDEO LINK

For more on the uses of transistors, watch the video at www.brightredbooks.net/N5Physics

DON'T FORGET

The transistor will switch on a device, or even a separate circuit, when its input voltage is above a given value.

Transistors

RELAY

A relay is an electromagnetic device which operates using a small current to switch on a circuit with a larger current.

The relay contains an electromagnet (a coil of wire wound around an iron core) which becomes magnetic when a current is in the coil. The electromagnet then closes a magnetic switch which completes a separate circuit.

Advantages of using a relay to switch on a separate circuit are:

- a small current can be used to safely switch on a separate circuit containing large currents or voltages

- several components (such as lamps) can be switched on and off simultaneously

- in circuits where sensors (like thermistors, LDRs or capacitors) provide input conditions for switching on, the relay can switch on components to alter the conditions (e.g. switching on a heater when a thermistor detects falling temperature).

relay coil　　**relay switch**
relay symbol

ONLINE TEST

Check how well you've learned about practical electrical and electronic circuits so far at www.brightredbooks.net/N5Physics

EXAMPLE

In the circuit shown, when the switch in a low voltage circuit is switched on, the relay coil is energised and switches on an external circuit using mains voltage to three lamps.

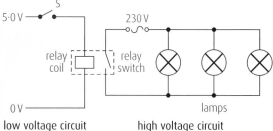
low voltage circuit　　high voltage circuit

THINGS TO DO AND THINK ABOUT

1. Find out how transistors are used as building blocks in the microchips of all electronic devices from computers to mobile phones.

PRACTICAL ELECTRICAL AND ELECTRONIC CIRCUITS: STANDARD ELECTRICAL COMPONENTS 2

A diode symbol

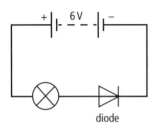

When connected in this way, the diode would conduct and the lamp would light

DIODES

Diodes are electronic devices which are used in circuits to control the direction of current. Diodes only conduct if they are connected the correct way round in a circuit, as shown in the diagram.

If you think of the diode symbol as looking like an arrowhead, then, for the diode (and circuit) to conduct, it should point towards the negative terminal of the supply.

EXAMPLE: 1

In this circuit, explain which lamps would light.

The diode connected to A is pointing the wrong way, so will not conduct – lamp A will be off.

Although one diode connected to B is pointing the correct way, the other diode is in series with it and is pointing the wrong way, so it will not conduct – lamp B will be off.

The diode connected to C is pointing the correct way and will conduct, so lamp C will light.

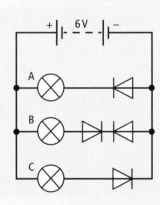

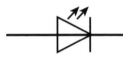

The symbol for a light emitting diode

An LED Screen TV

LIGHT EMITTING DIODE (LED)

Nowadays light emitting **diodes** are used in many different applications where an indicator light or a powerful light beam is required. LEDs use very little energy compared with conventional filament lamps which produce a lot of wasted heat energy. Also, LEDs have a longer life.

They are low voltage devices and commonly operate at a voltage of around 2 V and a current of around 20–200 mA. When used with a supply of more than 2 V, they are connected in series with a resistor, R, in a voltage divider circuit. This protects the LED from damage due to too large a voltage or current.

Remember, LEDs are diodes and only conduct if they are connected in circuits in the correct orientation.

When calculating the resistance of the resistor which protects the LED, information such as the operating current and voltage of the LED will be provided.

EXAMPLE 2

Calculate the resistance of R that allows the LED to operate at the correct voltage and current.

First calculate the voltage across resistor R.

$V_R = V_S - V_{LED} = 4 \cdot 5 - 1 \cdot 8 = 2 \cdot 7 \, V$

Then use Ohm's Law to calculate the resistance of R.

$R = \dfrac{V}{I} = \dfrac{2 \cdot 7}{15 \times 10^{-3}} = 180 \, \Omega$

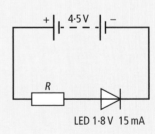

CAPACITORS

Capacitors are components which store charge.

Symbol for a capacitor

In Circuit 1, when switch S is closed, the capacitor begins to store charge. The voltage across the capacitor increases. The voltage at position X increases.

Facts about capacitors:

1. The capacitor becomes fully charged when the voltage across it gradually increases to reach the supply voltage.

2. At this point, the current in the circuit reduces to zero.

3. When a capacitor discharges, the voltage across it gradually decreases and reaches zero.

4. The length of time a capacitor takes to charge depends upon the capacitance of the capacitor and the resistance of the resistor in series with it. The greater the value of the capacitance or resistance, the longer the capacitor takes to charge up.

5. The unit for capacitance is the farad (F).

6. Typical values for capacitors are: 22 microfarads (22×10^{-6} F or $22\,\mu$F) or 15 picofarads (15×10^{-12} F or $15\,$pF).

DON'T FORGET

Work out the voltage across the resistor first, before using Ohm's Law.

Electrolytic capacitor

Capacitors are often used in timing circuits.

When the capacitor-charging part of Circuit 2 is connected to a transistor circuit, the voltage across the capacitor at position X becomes the input voltage to the transistor. When this voltage reaches the appropriate value of 0·7 V (but 2 V for a MOSFET transistor) the transistor will switch on (or conduct). In this circuit, the transistor switches on a buzzer, to make a countdown timer.

The time delay can be set by adjusting the value of the variable resistor which alters the time taken for the capacitor to charge.

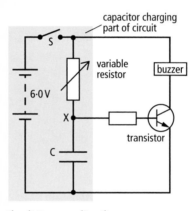

Circuit 2 – capacitor timer

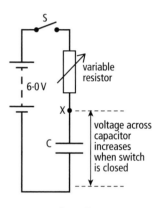

Circuit 1 – the voltage across a capacitor increases as more charge is stored

EXAMPLE 3

The capacitor circuit shown in Figure 1 is used in a car to delay an interior courtesy light switching off when the driver closes the door. The graph in Figure 2 shows the voltage across the capacitor from the time when the driver closes the door.

Explain how long the lamp remains on for after the driver closes the door.

The voltage across the capacitor decreases when the door is closed. After 9 seconds, the voltage across the capacitor falls to 0·7 V. The transistor stops conducting and the lamp switches off.

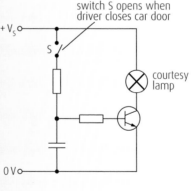

Figure 1

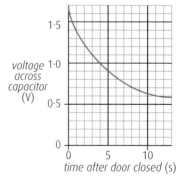

Figure 2

THINGS TO DO AND THINK ABOUT

1. Find out how the use of LEDs as light sources has reduced the energy bills in homes and in industry.

2. Find other applications which use capacitors in timing circuits.

DON'T FORGET

The greater the capacitance or resistance in a capacitor circuit, the longer the charging time.

PRACTICAL ELECTRICAL AND ELECTRONIC CIRCUITS: STANDARD ELECTRICAL COMPONENTS 3

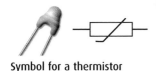

Symbol for a thermistor

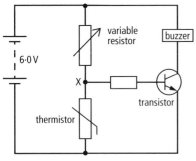

Circuit 3 – low-temperature warning circuit

THERMISTORS

Thermistors are resistors which change their resistance as their temperature changes. Most thermistors are designed to have a decrease in resistance as the temperature rises.

Circuit 3 shown left is used to switch on a warning buzzer whenever the temperature falls below a certain value.

When the temperature falls, the thermistor's resistance increases and so it gains a bigger share of the supply voltage. The voltage across the thermistor will eventually become bigger than 0·7 V at position X, causing the transistor to switch on and the buzzer to sound.

The resistance of the variable resistor can be adjusted, which changes the thermistor resistance required to switch on the transistor – giving a different switching temperature.

EXAMPLE 1

A thermistor is used as a temperature sensor in the circuit shown to monitor and prevent the air temperature in an office from exceeding safety levels.

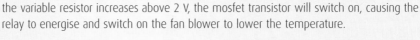

Explain how the circuit operates to prevent overheating.

As the office temperature increases, the thermistor's resistance decreases so it gains a smaller share of the supply voltage, causing the voltage across the variable resistor R to increase. When the voltage across the variable resistor increases above 2 V, the mosfet transistor will switch on, causing the relay to energise and switch on the fan blower to lower the temperature.

Explain how the variable resistor can be adjusted to make the fan start at a higher temperature.

At higher temperature, the resistance of the thermistor will be smaller and will obtain a smaller share of the supply voltage. The resistance of the variable resistor must be increased to obtain a larger share of the voltage.

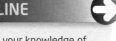

The symbol for an LDR

LIGHT DEPENDENT RESISTOR (LDR)

Light dependent resistors are devices which change their resistance as the light level changes; the resistance of LDRs decreases as the light level increases.

This light dependent resistor circuit shown here is used to switch on a high intensity LED whenever the daylight level falls below a fixed value.

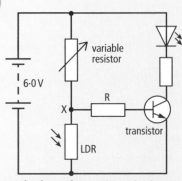

When the light level falls, the LDR's resistance increases and it gains a bigger share of the supply voltage. The voltage across the LDR will eventually become bigger than 0·7 V, and the transistor will switch on and the LED will light.

Light dependent resistor circuit

contd

The resistance of the variable resistor can be adjusted which will change the LDR resistance required to switch on the transistor – giving a different switching light level.

VIDEO LINK

Have a look at the LDR circuit clip at www.brightredbooks.net/N5Physics

EXAMPLE 2

A circuit that operates a motor to close a blind in a shop when the outside light level gets too high is shown.

Explain how the circuit operates to close the blind when the outside light level is too high.

The resistance of the LDR decreases as the light level increases so it gains a smaller share of the supply voltage, causing the voltage across the variable resistor to increase. When the voltage across the variable resistor increases above 0·7 V, the transistor will conduct causing the motor to switch on and operate the blind.

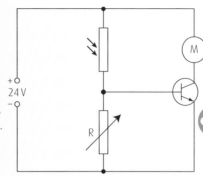

DON'T FORGET

This might help you remember how the resistance changes with temperature or light level:
• Temperature Up-Resistance Down (TURD)
• Light Up-Resistance Down (LURD)!

SOLAR CELL (PHOTOVOLTAIC CELL)

A solar cell transforms light energy into electrical energy. A voltage is produced when light shines on a solar cell.

solar cell

Symbol for a solar cell

Solar cells are commonly used to recharge batteries in portable or remote appliances (for example, your calculator might have a solar cell to recharge an internal battery).

EXAMPLE:

High intensity LEDs are used as path lights. The lights turn on automatically when it gets dark. The light contains a solar cell which charges a rechargeable battery during daylight hours. When dark, at night, the rechargeable batteries provide energy for the LEDs to light up the path.

At a particular daylight level, the voltage across the 400 Ω resistor is 0·4 V.

(a) Calculate the voltage across the rechargeable battery at this light level.

$$\frac{V_1}{V_2} = \frac{R_1}{R_2}$$

$$\frac{0·4}{V_2} = \frac{400}{1200}$$

$$V_2 = \frac{0·4 \times 1200}{400} = 1·2V$$

(b) Calculate the voltage generated by the solar cell.

$$V_{solar\ cell} = V_1 + V_2 = 0·4 + 1·2 = 1·6V$$

The solar cell and rechargeable batteries are part of the circuit shown here.

400 Ω

solar cell

1200 Ω

rechargeable batteries

This house has a bank of solar cells on the roof

A sunken LED light

THINGS TO DO AND THINK ABOUT

Find out how solar cells are being fitted to houses to help reduce electricity bills, and reduce the amount of electrical energy required from burning fossil fuels in Scotland.

DON'T FORGET

The greater the light intensity, the greater the voltage generated from the solar cell.

PRACTICAL ELECTRICAL AND ELECTRONIC CIRCUITS: RULES FOR CURRENT AND VOLTAGE IN SERIES AND PARALLEL CIRCUITS

The key concept to learn in this topic is:

- application of the rules for current and potential difference (voltage) in series and parallel circuits
  $I_s = I_1 = I_2 = ...,\ V_s = V_1 + V_2 + ...,\ I_p = I_1 + I_2 + ...,\ V_p = V_1 = V_2 = ...$

CURRENT AND VOLTAGE RELATIONSHIPS IN A SERIES CIRCUIT

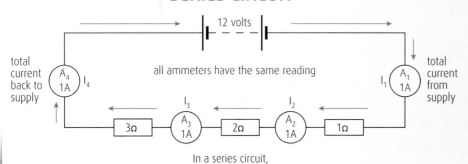

In a series circuit,
$I_s = I_1 = I_2 = ...$

Current

In a series circuit, the same value of current is measured at all positions in the circuit.

In this circuit, the value of the current measured by all four ammeters is the same.

DON'T FORGET

In series circuits, the voltages across each of the components add up to the supply voltage.

EXAMPLE 1

State the current readings on A_2, A_3 and A_4 in the following circuit:

This is a series circuit. This means that the current has the same value at all points in the circuit.

So $A_1 = A_2 = A_3 = A_4 = 2A$

Voltage

In a series circuit, the supply voltage divides across each component.

The voltages across each of the components adds up to the supply voltage.

In a series circuit, $V_s = V_1 + V_2 + ...$

$V_1 + V_2 + V_3 = 1.5 + 4.0 + 0.5 = 6V$

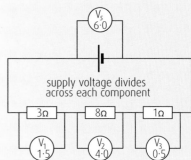

supply voltage divides across each component

DON'T FORGET

The greatest value of resistance receives the greatest share of the voltage.

EXAMPLE 2

State the voltage readings on V_2, V_3, V_4, and V_5.

The supply voltage divides across each resistor. As each resistor has the same value, the supply voltage will divide equally between them.

So

$V_2 = V_3 = V_4 = V_5 = 4V$

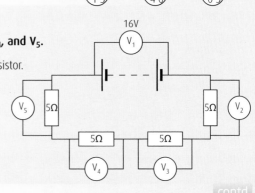

contd

A series circuit containing lamps can be used as a decoration on a Christmas tree.

Each of the lamps operates at 5 V. The circuit is connected to the 230 V mains supply.

EXAMPLE 3

Calculate how many lamps are required in this circuit to allow them to operate at their suggested operating voltage.

The lamps are identical and so have the same value of resistance.

This means that the supply voltage will divide equally across all lamps.

Mains voltage = 230 V; voltage of one lamp = 5 V.

$\frac{230}{5} = 46$

Therefore 46 lamps are required to operate each lamp at 5 V.

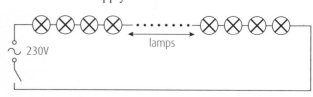

CURRENT AND VOLTAGE RELATIONSHIPS IN A PARALLEL CIRCUIT

Current

In a parallel circuit, the supply current **splits up** through each component when it reaches a branch in the circuit. In this circuit, the total 9 A supply current splits through each branch of the circuit.

$$A_1 + A_2 + A_3 = 6 + 2 + 1 = 9\,A$$

Voltage

In a parallel circuit, the voltage across components connected in parallel **remains the same** as the supply voltage.

The supply voltage of 6 V is measured across each resistor connected in parallel with the supply.

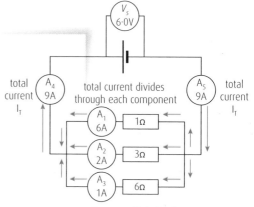

In a parallel circuit,
$I_p = I_1 + I_2 + ...$

EXAMPLE 4

The three lamps in the circuit shown are identical.

The reading on the ammeter is 1·2 A when the lamps operate at the correct voltage.

Calculate the current in one lamp.

$I = \frac{1\cdot2}{3} = 0\cdot4\,A$

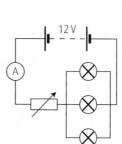

EXAMPLE 5

Ceiling lamps are connected to the mains in the circuit shown.

Each lamp is rated at 230 V, 20 W.

Explain why the lamps must be connected in parallel.

Lamps must operate at 230 V. Lamps in parallel receive the supply voltage (230 V)

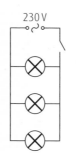

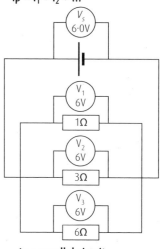

In a parallel circuit,
$V_p = V_1 = V_2 = ...$

ONLINE

To see the differences between series and parallel circuits. In practice, watch the video at www.brightredbooks.net/subjects/N5physics

THINGS TO DO AND THINK ABOUT

In the household wiring, devices which operate at 230 V a.c. are plugged into sockets and work normally. Investigate how the 'ring mains' circuit is used in household wiring circuits to ensure that the devices operate at the correct working voltage.

PRACTICAL ELECTRICAL AND ELECTRONIC CIRCUITS: TOTAL RESISTANCE IN SERIES, PARALLEL AND COMBINATIONS OF SERIES AND PARALLEL CIRCUITS

The key concepts to learn in this topic are:

- knowledge of the effect on the total resistance of a circuit of adding further resistance in series or in parallel
- the use of appropriate relationships to solve problems involving the total resistance of resistors in series and in parallel circuits, and circuits with a combination of series and parallel resistors:

$$R_T = R_1 + R_2 + ..., \quad \frac{1}{R_T} = \frac{1}{R_1} + \frac{1}{R_2} + ...$$

DON'T FORGET

When the total *resistance* of a circuit *increases*, the supply current *decreases*.

ohmmeter ≡ ohmmeter

DON'T FORGET

When the resistance values have mixed units (like ohms Ω and kilohms kΩ) they must be converted into ohms before using the relationship.

VIDEO LINK

Check out the tutorial for more on resistors in series at www.brightredbooks.net/N5Physics

RESISTORS IN SERIES CIRCUITS

When resistors are connected in *series*, the total resistance of the circuit *increases*.

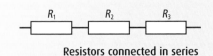

Resistors connected in series

One single resistor can replace several resistors connected in series.

For example, a single resistor of value $17\,k\Omega$ can replace three resistors of values $1\,k\Omega$, $4\,k\Omega$ and $12\,k\Omega$ which are connected in series in a circuit.

Add to get the sum of resistors in a series circuit.

The combined resistance of resistors connected in series can be found using the relationship $R_T = R_1 + R_2 + ...$:

Note that in a circuit where resistors are connected in series, if a further resistor is added in series, then the total resistance of the circuit increases.

EXAMPLE 1

Calculate the total resistance in circuit in the diagram.

$R_T = R_1 + R_2 + R_3$
$\quad = 1{\cdot}2 \times 10^3 + 250 + 15 \times 10^3 = 16450\,\Omega$

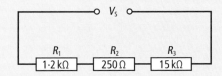

DON'T FORGET

The combined resistance of resistors in parallel is always less than the resistance of the smallest resistor.

RESISTORS IN PARALLEL CIRCUITS

When more resistors are connected in parallel, the total resistance of the circuit *decreases*.

One single resistor can replace several resistors connected in parallel in a circuit. For example, a single resistor of value $1{\cdot}2\,k\Omega$ can replace three resistors of value $2\,k\Omega$, $4\,k\Omega$ and $12\,k\Omega$ that are connected in parallel in a circuit.

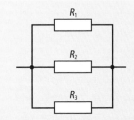

Resistors connected in parallel

The combined resistance of resistors connected in parallel can be found using the relationship:

$$\frac{1}{R_T} = \frac{1}{R_1} + \frac{1}{R_2} + ...$$

Note that in a circuit where resistors are connected in parallel, if a further resistor is added in parallel, then the total resistance of the circuit decreases.

contd

EXAMPLE 2

Calculate the combined resistance in the circuit in the diagram.

$$\frac{1}{R_T} = \frac{1}{R_1} + \frac{1}{R_2} + ...$$

$$\frac{1}{R_T} = \frac{1}{R_1} + \frac{1}{R_2} + \frac{1}{R_3}$$ Use this relationship to calculate the answer.

$$= \frac{1}{3} + \frac{1}{5} + \frac{1}{12}$$ Write out the fractions.

$$= 0.333 + 0.2 + 0.083 = 0.616$$ Calculate the fractions (usually to three decimal places).

$$R_T = \frac{1}{0.616}$$ Calculate 1 divided by this intermediate answer

$$= 1.62 k\Omega$$ to get the final answer.

Since all resistance values in the example are given in kilohms, there is no need to convert into ohms. The units for the final answer will be kilohms.

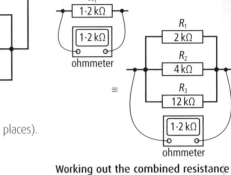

Working out the combined resistance of resistors in parallel

When resistors connected in parallel have the same resistance values, there is a shortcut to calculate the combined resistance instead of using the relationship. The combined resistance of three identical resistors is $\frac{1}{3}$ of the single resistance. If there are two identical resistors, their combined resistance is $\frac{1}{2}$ of the single resistance.

EXAMPLE 3

Use the shortcut to calculate the combined resistance of the resistors in the circuit.

Total number of resistors = 3

Total resistance = $\frac{1}{3}$ of one resistance = $\frac{12}{3}$ $R_T = 4\Omega$

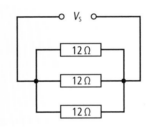

> **DON'T FORGET**
>
> Questions which require calculations using this relationship are very likely to feature in the exam so make sure you try enough examples to understand it really well.

> **DON'T FORGET**
>
> In complex resistor circuits try to recognise and separate series and parallel parts.

RESISTORS IN COMPLEX CIRCUITS

Complex circuits have combinations of resistors connected in both series parts and parallel, as shown in the diagram. When circuits contain a mixture of resistors connected in series and in parallel, to calculate the total resistance, both relationships

$$R_T = R_1 + R_2 ...   \text{and}   \frac{1}{R_T} = \frac{1}{R_1} + \frac{1}{R_2} + ...   \text{have to be applied separately.}$$

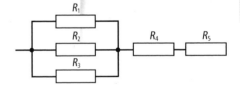

Complex combination of resistors

EXAMPLE 4

Calculate the total resistance in this circuit between points X and Y.

Part of the circuit has resistors connected in series, and part has resistors connected in parallel. It is important to separate the two parts.

To find the combined resistance of the circuit, the total resistance of each part must be calculated separately.

Part 1: $R_T = 17 k\Omega$ (See Example of resistors connected in series.)

Part 2: $R_T = 1.62 k\Omega$ (See Example 2 for calculation of resistors connected in parallel.)

Since Part 1 and Part 2 are in series, the final *total* resistance is the sum of these two resistance values.

$$R_T = R_1 + R_2 = 17 + 1.62 = 18.62 \ k\Omega$$

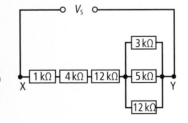

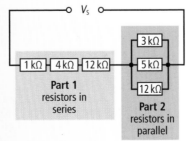

Work out which resistors are in series and which are in parallel

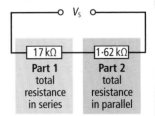

Part 1 of the circuit is in series with part 2

> **ONLINE TEST**
>
> Test your knowledge of current, voltage and resistance online, at www.brightredbooks.net/N5Physics

ELECTRICAL POWER

The key concepts you will learn in this topic are:

- definition of electrical power in terms of electrical energy and time
- the use of an appropriate relationship to solve problems involving energy, power and time:
  $$P = \frac{E}{t}$$
- knowledge of the effect of potential difference (voltage) and resistance on the current in and power developed across components in a circuit

- the use of an appropriate relationship to solve problems involving power, potential difference (voltage), current and resistance in electrical circuits:
  $$P = IV, \; P = I^2R, \; P = \frac{V^2}{R}$$
- the selection of an appropriate fuse rating given the power rating of an electrical appliance. (3 A fuse for most appliances rated up to 720 W, 13 A fuse for appliances rated over 720 W.)

VAC

Model: V-015T

220–240 V - 50 Hz
1600 W

Serial number: ☐ CE
01463/33 Made in China

Appliance rating plates tell you the power rating

POWER

When energy is transferred from one store to another, for example when water stored behind a dam (E_p) flows down pipes (E_k) to a river below, it takes time for the energy to transfer.

Whenever there is a current in a circuit, electrical energy is transferred. The electrical energy can be transferred into light (by a lamp), or heat (by a heater), or kinetic energy (by a motor) depending on the appliance in the circuit.

The rate at which energy is transferred is known as *power*. Electrical power is defined as the rate at which electrical energy is transferred by an electric circuit. Power is measured in joules per second (Js^{-1}), or watts (W). The relationship between electrical power, P, energy, E, and time, t, is $P = \frac{E}{t}$.

Power can be calculated in a number of ways:

1. $P = \dfrac{E}{t}$ where:

   P is the power in watts (W)
   E is the energy in joules (J)
   t is the time in seconds (s)

2. $P = IV$ where:

   P is the power in watts (W)
   I is the current in amperes (A)
   V is the voltage in volts (V)

3. $P = I^2R$ where:

   P is the power in watts (W)
   I is the current in amperes (A)
   R is the resistance in ohms (Ω)

4. $P = \dfrac{V^2}{R}$ where:

   P is the power in watts (W)
   V is the voltage in volts (V)
   R is the resitatnce in ohms (Ω)

EXAMPLE 1

A 3 kW electric fire is switched on for 2½ hours. Calculate the energy used.

$P = \dfrac{E}{t}$ (convert 2½ hours into seconds) $E = 3000 \times 2{\cdot}5 \times 60 \times 60 = 27\,000\,000\,J$

$3000 = \dfrac{E}{2{\cdot}5 \times 60 \times 60}$

EXAMPLE 2

A 3 kW electric fire is switched on for 2½ hours. Calculate the number of kilowatt-hours (kWh) used.

Kilowatt-hours are an alternative unit for energy, often used by power companies. To calculate kilowatt-hours, use the same formula but keep the units for power and time in kilowatts and hours.

$P = \dfrac{E}{t}$ so $3 = \dfrac{E}{2{\cdot}5}$ $E = 3 \times 2{\cdot}5 = 7{\cdot}5\,kWh$

contd

EXAMPLE 3

Calculate the power rating of a 3·2 V ultra-bright LED which has an operating current of 30 mA.

$P = IV = 30 \times 10^{-3} \times 3·2 = 0·01\,W$

EXAMPLE 4

In a 250 km length of overhead power cable, the power loss due to the heating effect of current in the wires is 14 MW. The current in the cables is 900 A. Calculate the resistance of the cables.

$P = I^2R$

$14 \times 10^6 = 900^2 \times R$

$R = \dfrac{14 \times 10^6}{900^2} = 17·3\ \Omega$

EXAMPLE 5

Calculate the potential difference across the cable in Example 4.

$P = IV$

$14 \times 10^6 = 900 \times V$

$V = \dfrac{14 \times 10^6}{900} = 1·6 \times 10^4\,V$

EXAMPLE 6

A 1·2 kW mains-operated hairdryer has a current of 0·04 A when blowing cold air only. Calculate the resistance of the heating element.

Power of motor only (cold air)

$P = IV = 0·04 \times 230 = 9·2\,W$

1·2 kW = 1200W This is the total power of the hairdryer when both motor and heater are working.

Power of heating element only = 1200 – 9·2 = 1190·8 W

Note that, since the appliance is mains operated V = 230 V

$P = \dfrac{V^2}{R}$

$1190·8 = \dfrac{230^2}{R}$

$R = \dfrac{230^2}{1190·8} = 44·4\ \Omega$

EFFECT OF VOLTAGE AND RESISTANCE ON CURRENT IN AND POWER DEVELOPED IN A COMPONENT

The current in an electrical component and the power developed in it depend on the resistance and the voltage across the component. Consider a resistor, R, with a voltage, V, applied across it.

The equation used to calculate electrical power is: $P = \dfrac{V^2}{R}$

Consider the following circuits where the voltage and resistance are increased separately

If voltage increases:

- Power increases
- Current increases

$V\uparrow \begin{smallmatrix} P\uparrow \\ I\uparrow \end{smallmatrix}$

If resistance increases:

- Power decreases
- Current decreases

$R\uparrow \begin{smallmatrix} P\downarrow \\ I\downarrow \end{smallmatrix}$

The current in an electrical component and the power developed in it depends on the resistance and the voltage across the component.

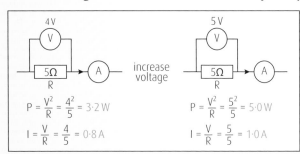

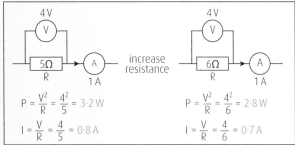

FUSE RATINGS

The current rating (e.g. 3 A or 13 A) of fuses in plugs of domestic appliances is determined by the Power Rating of the appliance.

For most appliances with a power rating up to 720 W a 3 A fuse should be used.
For appliances with a power rating greater than 720 W a 13 A fuse should be used.

SPECIFIC HEAT CAPACITY 1

The key concepts to learn in this topic are:

- knowledge that different materials require different quantities of heat to raise the temperature of unit mass by one degree Celsius
- the use of an appropriate relationship to solve problems involving mass, heat energy, temperature change and specific heat capacity:

$$E_h = cm\Delta T$$

- knowledge that the temperature of a substance is a measure of the mean kinetic energy of its particles
- the use of the principle of conservation of energy to determine heat transfer.

VIDEO LINK

Head online and watch the video of a hot iron bar and see how different the particles look when heated at www.brightredbooks.net/N5Physics

DON'T FORGET

When substances are heated and their temperature rises, this is because the average kinetic energy of all the particles in the substance has increased.

Far more heat energy is required to heat a bath full of water to body temperature than is needed to heat water for a cup of tea to a far higher temperature.

DON'T FORGET

The change in temperature for a material is written as ΔT where the symbol Δ (pronounced 'delta') stands for the change in temperature of the material.

DON'T FORGET

1 kg of copper will heat up at a different rate than 1 kg of water.

MEAN KINETIC ENERGY OF PARTICLES OF A SUBSTANCE AND TEMPERATURE

The **atoms** and **molecules** which make up every substance are continually moving at different **speeds** and with different kinetic energies. The temperature of a substance is proportional to the mean or average value of these kinetic energies.

When heat energy is transferred to these atoms and molecules, they gain more kinetic energy.

When a substance is heated and its temperature rises, this is because the average value of the kinetic energy of all of the atoms or molecules has increased.

particles vibrate and bump into neighbouring particles in the object

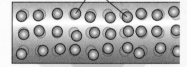

The effect of heating on the particles in an object

TEMPERATURE CHANGE AND HEAT ENERGY

To be able to determine accurately the amount of heat energy required to change the temperature of different materials a detailed study is needed.

When heat energy is added to or removed from a material, causing its temperature to rise or fall, the *change* in temperature depends on:

- the type of material
- the mass of the material
- the amount of heat energy added or removed.

The effect of changing each of these quantities is shown in the following experiments.

Experiment 1 - heating different materials

In the following experiment, equal masses of aluminium, water and iron were given the same amount of heat energy and the temperature change (ΔT) was recorded in each case. The results are shown in the table.

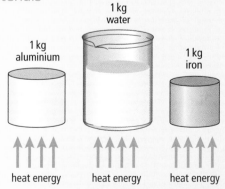

Material	Heat energy added (J)	Temperature change ΔT (°C)
aluminium	37 500	42
water	37 500	9
iron	37 500	78

contd

The results show that when the same amount of heat energy is added to, or removed from, the same mass of different materials, the temperature change is different. The temperature change, ΔT, depends on a quantity called the specific heat capacity for the material (given the symbol, c).

The units for specific heat capacity are $J\,kg^{-1}\,^{\circ}C^{-1}$.

Experiment 2 – adding different amounts of heat energy to equal masses of the same material

The same mass of water is heated in identical beakers. Different amounts of heat energy are added to each beaker and the temperature change, ΔT, recorded. The results are shown in the table.

Heat energy added (J)	Temperature change ΔT (°C)
37 500	9
75 000	18
102 500	27

The results show that when equal masses of the same material are given different amounts of heat energy, the temperature change, ΔT, is different. Temperature change depends on the energy supplied; the greater the heat energy supplied, the larger the temperature change.

Experiment 3 – adding equal amounts of heat energy to different masses of the same material

The same amount of heat energy is added to different masses of aluminium, and the temperature change, ΔT, is recorded. The results are shown in the table.

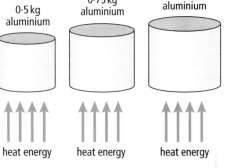

Mass of aluminium (kg)	Temperature change ΔT (°C)
0·5	48
0·75	36
1·0	24

The results show that when different masses of the same material are given equal amounts of heat energy, the temperature change, ΔT, is different. Temperature change depends on the mass of the material. The smaller mass has a greater temperature change.

These experiments show that the for the equal masses (or unit mass) of different materials, different amounts of heat energy are required to cause the same temperature change in each material. This can be summarised as:

'Different materials require different quantities of heat energy to raise the temperature of unit mass by one degree Celsius.'

 THINGS TO DO AND THINK ABOUT

Investigation

1. When two separate masses of water at different starting temperatures are mixed, the *final* temperature of the mixture lies between the two starting temperatures. How would you calculate the final temperature of the mixture?

2. Think about what happens to the heat energy of each liquid after they have mixed. Which one loses heat energy and which one gains heat energy?

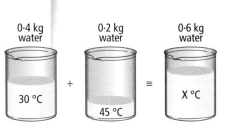

3. What can you say about the heat energy lost by one mass and gained by the other? Could you use this information to write an equation which could be used to calculate the final temperature?

4. *Safely* prepare some masses of water at different temperatures.

5. Calculate the final temperature of the mixture, then *carefully* do the mixing to see if the final measured temperature agrees with your calculation.

SPECIFIC HEAT CAPACITY 2

DON'T FORGET

The value for c for the substance can be obtained from the Data sheet or may sometimes be given in the question.

RELATIONSHIP FOR SPECIFIC HEAT CAPACITY

E_h, m, c, and ΔT are used together in a single relationship when heat energy, added or removed from a substance, causes a temperature change:

$E_h = cm\Delta T$ where: E_h is the heat energy added or removed (J)
c is the specific heat capacity of the substance ($J\,kg^{-1}°C^{-1}$).
m is the mass of the substance (kg).
ΔT is the temperature change of the substance (°C).

DON'T FORGET

In calculations, masses given in grams must be converted into kilograms.

EXAMPLE 1

(a) How much heat energy is required to raise the temperature of 840 g of water on a cooker from 21°C to 84°C?

c for water = 4180J kg^{-1} °C^{-1} (from the Data sheet)

First calculate ΔT = 84 −21 = 63°C Convert 840g into 0·84kg

$E_h = cm\Delta T$
   $= 4180 \times 0·84 \times 63$
   $= 221206J$

(b) In practice, the actual amount of energy supplied by the cooker to cause this temperature rise is greater. Give one reason for this difference.

As the water is being heated, some heat will be lost to the surroundings, so more heat energy will be required from the cooker.

DON'T FORGET

Example 1b is a common question. Heat lost to the surroundings is a common answer to explain the difference between calculated and actual heat energy required.

EXAMPLE 2

When 4·4 × 10⁴ J of heat energy is added to a dinner plate in an oven, its temperature rises from 18°C to 87°C . The specific heat capacity of the dinner plate is 2700 J kg⁻¹°C⁻¹. Calculate the mass of the plate.

ΔT = 87 − 18 = 69°C

$E_h = cm\Delta T$

$4·4 \times 10^4 = 2700 \times m \times 69$ so $m = \dfrac{4·4 \times 10^4}{2700 \times 69} = 0·24kg$

The Principle of Conservation of Energy

When energy is transformed or converted from one form of energy into another, the total amount of energy does not change. For example, when electrical energy is converted into heat energy by a water heater, there is no loss of energy.

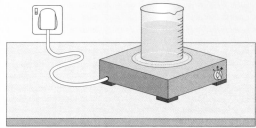

Energy is neither created nor destroyed during the process.

This is known as the Principle of Conservation of Energy.

In practice, not all of the electrical energy transformed into heat energy by the electric heater will heat the water. Some heat energy will heat up the actual heater itself, some heat energy may heat up the surrounding air (this is known as 'energy loss' to the surroundings).

Sometimes reducing this energy loss may involve adding insulation (or even a lid) to a container being heated.

DON'T FORGET

Exam questions about heat energy are usually quite complex and may involve more than simple calculations. They often start by describing a particular situation or application. They usually require some of the data needed for use in the equation $E_h = cm\Delta T$ to be calculated. There is often a further part of the question which requires you to think about the answer and then comment on the result.

contd

EXAMPLE 3

(a) An oil-filled radiator has an electric heater inside it to raise the temperature of 12·5 kg of oil to an operating temperature of 128°C. Calculate the energy required to raise the temperature of this oil from 20°C. The specific heat capacity of the radiator oil is 2200 J kg⁻¹°C⁻¹.

$E_h = cm\Delta T$ $\qquad\qquad \Delta T = 128 - 20 = 108\,°C$
$\quad = 2200 \times 12\cdot5 \times 108 = 2\,970\,000\,J$

(b) The radiator heater is rated at 1500 W. Calculate the minimum time for the oil to reach its operating temperature.

$E = P \times t$
$2\,970\,000 = 1500 \times t \quad so \quad t = \dfrac{2\,970\,000}{1500} = 1980\,s$

(c) In practice, the actual time taken is greater than the minimum time calculated. Give one reason for this difference.

As the radiator is heated, some heat energy will be lost to the surroundings. More heat energy will be required to raise the temperature.

EXAMPLE 4

Disc brakes are used on mountain bikes to slow down quickly. They are usually made of iron which is quite heavy. When the brakes are applied, the kinetic energy of the bike and cyclist are converted into heat energy in the metal discs.

A manufacturer decided to reduce the **weight** of the bike by using aluminium disc brakes instead of iron. On the new model, the total mass of the aluminium disc brakes is 1·25 kg.

While testing this new bike, a cyclist travels down a steep hill using the brakes to slow down. During this braking, 676 500 J of kinetic energy is converted into heat energy in the disc brakes. Before braking, the temperature of the disc brakes is 25°C. Assume that all of the kinetic energy is converted into heat energy in the disc brakes.

(a) Calculate the final temperature of the brakes.

$E_h = cm\Delta T$

Specific heat capacity of aluminium = 902 J kg⁻¹°C⁻¹ (obtained from the Data sheet)

$676\,500 = 902 \times 1\cdot25 \times \Delta T$

$\Delta T = \dfrac{676\,500}{902 \times 1\cdot25}$

$\quad = 600\,°C$

Add the starting temperature to get the final temperature

$25 + 600 = 625°C$

(b) Explain why the final temperature is likely to be slightly less than that calculated above.

Some of the heat energy produced will be lost to the surroundings as the cyclist travels down the hill. This will reduce the final temperature.

(c) Use the Data sheet to comment on why the manufacturer decided that aluminium is unsuitable as a material for the disc brakes compared with iron.

The melting point of aluminium is 660°C (from the Data sheet). The final temperature of the aluminium is close to its melting point. The metal would become dangerously soft.

THINGS TO DO AND THINK ABOUT

Investigation

1. Some kettles have markers to tell you when there is enough water for one cup. What would you need to know to find out the difference in the electricity cost if the kettle is filled:

   • to the maximum level   • to one cupful level.

   (Hint: you need to find the *difference* in heat energy needed to heat one cupful and one full kettle to boiling from, say, 20°C.)

DON'T FORGET

The Data sheet in the exam may be required to obtain information not given in the question. Be familiar with each table in the sheet.

ONLINE TEST

For more questions on specific heat capacity at test yourself at www. brightredbooks.net/ N5Physics

ONLINE

Investigate more about specific heat capacity at www.brightredbooks.net/ N5Physics

DON'T FORGET

• Remember that heat energy can be added to or removed from a substance, and a temperature change can be an increase or decrease.
• Always select the value for c from the Data sheet, unless it is given in the question.
• When calculating the heat energy in specific heat capacity questions, it may be necessary to use one of a number of expressions for energy:
$E = P \times t$ $\quad E = (VI)t$
$E = F \times d$ $\quad E = \frac{1}{2}mv^2$
• The final parts of questions concerning heat energy often ask about improving an experiment, for example, how to reduce the amount of heat loss. Think carefully about the answer, which may be connected with insulating the container. (The answer can be as simple as putting a lid on a container!)

SPECIFIC LATENT HEAT

The key concepts to learn in this topic are:

- knowledge that different materials require different quantities of heat to change the state of unit mass
- knowledge that the same material requires different quantities of heat to change the state of unit mass from solid to liquid (fusion)

and to change the state of unit mass from liquid to gas (vaporisation)

- the use of an appropriate relationship to solve problems involving mass, heat energy and specific latent heat:

$E_h = ml$

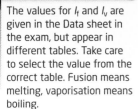

1 kg ice at 0°C

latent heat energy added, ice changes to water

latent heat energy removed, water changes to ice

1 kg water at 0°C

1 kg water at 100°C

latent heat energy added, water changes to steam

latent heat energy removed, steam changes to water

1 kg steam at 100°C

Latent heat and changes of state

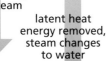

CHANGE OF STATE

There are three states of matter – solid, liquid and gas.

The change of state from solid to liquid is called **fusion** or *melting*, and from liquid back to solid is called *solidifying* or *freezing*. The change of state from liquid to gas is called **vaporisation**, and gas to liquid is called **condensation**.

There is *no change in temperature* when a change of state takes place. So, when water at 100°C changes into steam at 100°C, the temperature remains constant. When heat energy is added to ice at 0°C, the ice eventually changes into the same mass of water, still at 0°C.

A solid substance at its melting point contains less heat energy than an equal mass of the same liquid substance at the melting point. The difference in energy is due to latent heat energy.

When a substance changes state, this happens at the *melting point* (in the case of fusion) or the *boiling point* (in the case of vaporisation) of the substance. The temperature remains constant.

Latent heat energy can be added or removed to cause the change of state. The energy required to change the state of a substance depends on the *mass* and *type* of substance and whether the change causes melting or vaporisation.

Different amounts of heat energy are required to change the state of equal masses (or unit mass) of different materials. This can be summarised as:

'Different materials require different quantities of heat energy to change the state of unit mass.'

The energy required to change the state of 1 kg of a substance from solid to liquid is called the specific latent heat of fusion of the substance (measured in $J kg^{-1}$).

The energy required to change the state of 1 kg of a substance from liquid to gas is called the specific latent heat of vaporisation of the substance (measured in $J kg^{-1}$).

Also, the same material requires different quantities of heat energy to change the state of the same mass (or unit mass) from solid to liquid (fusion) and to change the state of unit mass from liquid to gas (vaporisation).

- l_f is the specific latent heat of fusion of a substance.
- l_v is the specific latent heat of vaporisation of a substance.

The relationship used to calculate the heat energy required to change the state of a substance of mass m is:

$E_h = ml_f$ for a substance at its melting point

$E_h = ml_v$ for a substance at its boiling point

There is no need for ΔT in these relationships as the temperature does not change.

EXAMPLE 1

Calculate the heat energy required to change 2·4 kg of solid iron into liquid iron.

The specific latent heat of fusion of iron = $2·67 \times 10^5 J kg^{-1}$ (from the Data sheet).
$E = ml_f = 2·4 \times 2·67 \times 10^5 = 640800 J$

contd

EXAMPLE 2

A kettle with a power rating of 2·5 kW is filled with 1·2 kg of water and switched on by a student. When the kettle reaches boiling point, the automatic cut-off switch fails to operate and the kettle continues to heat the water for 4 minutes before it is switched off manually.

(a) Calculate the heat energy supplied by the kettle during the 4 minutes.

$E = Pt$ $\quad\quad$ ($P = 2·5$ kW $= 2500$ W, $t = 4$ minutes $= 4 \times 60 = 240$ s)

$E = 2500 \times 240 = 6 \times 10^5$ J

(b) Calculate the mass of water remaining in the kettle after it is switched off.

The specific latent heat of vaporisation of water $= 22·6 \times 10^5$ J kg^{-1} (from the Data Sheet)

$E = ml_v$

$6 \times 10^5 = m \times 22·6 \times 10^5$ $\quad$ so $\quad m = 0·27$ kg

So mass of water remaining in kettle $= 1·2 - 0·27 = 0·93$ kg

(c) Explain why this value is likely to be inaccurate.

While heating the water in the kettle, some heat energy will be lost to the surroundings instead of vaporising the water.

ONLINE TEST

For more questions on specific latent heat, test yourself at www.brightredbooks.net/N5Physics

EXAMPLE 3

A solid of mass 125 g in a container is heated using a 50 W hotplate heater.

The graph shows how the temperature of the solid changes as it is heated from 25°C to 95°C.

(a) What is the melting point of the solid? Explain your answer.

The melting point is 50°C. The temperature of the solid remains constant at 50°C while it changes state to liquid.

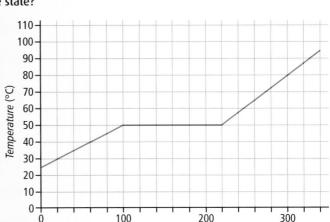

substance

temperature probe

hotplate

(b) How long does it take for the solid to change state?

120 seconds.

(c) How much heat energy was supplied to the solid while it changed state?

Power of heater $= P = 50$ W
Time taken to melt solid $= t = 120$ s

$E = Pt = 50 \times 120 = 6000$ J

(d) Calculate the specific latent heat of fusion of the solid.

Energy required to melt solid $= 6000$ J
Mass of solid $= 125$ g $= 0·125$ kg

$E = ml_f$

$6000 = 0·125 \times l_f$

$l_f = 4·8 \times 10^4$ J kg^{-1}

 THINGS TO DO AND THINK ABOUT

1. Investigate how the energy transferred to ice when it melts is used to cool other objects down, e.g. ice packs applied to a sprained ankle.

2. Refrigerators keep food cool. Find out how the 'coolant liquid' used in fridges continually vaporises and condenses in order to transfer heat energy from food.

ONLINE

Check out the specific latent heat link at www.brightredbooks.net/N5Physics

GAS LAWS AND THE KINETIC MODEL 1

The key concepts to learn in this topic are:
- definition of pressure in terms of force and area
- the use of an appropriate relationship to solve problems involving pressure, force and area:

$$p = \frac{F}{A}$$

- description of how the kinetic model accounts for the pressure of a gas

Sinking into snow

Staying on surface of snow

DON'T FORGET +

Larger force means greater pressure; larger area means smaller pressure.

VIDEO LINK ▶

Head online and watch the clip showing how polystyrene cups are affected by strong water pressure at www.brightredbooks.net/N5Physics

DEFINING PRESSURE

Pressure is defined as force per unit area.

Pressure occurs when a force is applied to a surface. Whenever a force is applied to an object, pressure is exerted on the object. For example, the force can be caused by a hammer striking a nail, or by a molecule of a gas colliding with the walls of its container.

The relationship for pressure connects the force and the surface area of the object.

$$\text{pressure} = \frac{\text{force}}{\text{area}}$$
$$p = \frac{F}{A}$$

where p = pressure (Pa), F = force (N) and A = area (m^2)

The units for pressure are pascal (Pa) or newtons per square metre, (Nm^{-2}) and $1\,\text{Pa} = 1\,\text{Nm}^{-2}$.

The *larger* the force, the *greater* the pressure; the *larger* the area, the *smaller* the pressure. For example, a person standing on snow exerts a pressure on the snow through their feet because the feet are in contact with the ground. The area of the feet in contact with the snow is small, so the pressure on the snow is great and the person sinks into the snow.

If the person stands on skis, the area in contact with the snow is much greater, so the pressure on the snow is smaller and the person stays on the surface.

Other examples of situations in which pressure is important are:

- a sharp knife is used for cutting vegetables – the edge of the knife is very narrow, so a very small area is in contact with the vegetables; the pressure is very large and cutting is easier

- a camel has very large feet to spread the force of its weight over a large surface area – this prevents it sinking into soft sand in the desert

- farm tractors have very wide tyres to increase the area of each tyre that is in contact with the ground – this reduces the pressure of the heavy tractor on the soil.

EXAMPLE 1

A container of mass 4.0×10^4 kg rests on the trailer of a lorry.

(a) Calculate the force exerted by the container on the trailer.

The weight of the container causes a force on the trailer.

$F = W = mg = 4.0 \times 10^4 \times 9.8 = 3.92 \times 10^5\,\text{N}$

(b) The area of the container surface in contact with the trailer is 24.5 m^2. Calculate the pressure exerted on the trailer by the container.

$p = \frac{F}{A} = \frac{3.92 \times 10^5}{24.5} = 1.6 \times 10^4\,\text{Pa}$

contd

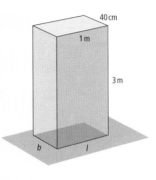

EXAMPLE 2

A concrete block of mass 200 kg has dimensions 3 m by 1 m by 40 cm. What is the greatest pressure that the block can exert on a flat horizontal surface?

The greatest pressure is exerted when the smallest face of the cube is in contact with the surface.

$$F = mg = 200 \times 9.8 = 1.96 \times 10^3 \, N$$

$$p = \frac{F}{A} = \frac{1.96 \times 10^3}{0.4} = 4.9 \times 10^3 \, Pa$$

DON'T FORGET

Area must always be given in square metres, so remember to convert centimetres or millimetres.

ONLINE TEST

For a test on pressure, visit www.brightredbooks.net/N5Physics

THE KINETIC MODEL OF A GAS

The 'kinetic model' describes gas pressure in terms of the behaviour and movement of the particles making up the gas. A container filled with any gas contains millions of tiny particles, continually and freely moving in random directions at high speeds. When these moving gas particles collide with each other, or with the walls of their container, no energy is lost.

GAS PRESSURE

The pressure exerted by a gas is caused by the moving particles that collide with the walls of the gas container. Each collision causes the wall to receive a tiny force. The size of the pressure depends on the *number of collisions* and the *average force per collision* exerted on the area of the walls. For example, if you cover the end of a bicycle pump with your thumb and press the plunger, it becomes very hard to keep your thumb in position because the gas pressure inside the pump increases. As the particles move, they collide with other particles and also with the inside walls of the pump. The pressure that you feel is due to the average force produced by millions of particle collisions with your thumb!

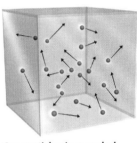

Gas particles in a sealed container

GAS VOLUME

The volume of a gas is the amount of space the particles have to move around in. If the walls of the container are brought closer, there will be less time between particle collisions with the wall. In the bicycle pump example, the volume is reduced when the plunger is pressed – hence more collisions per second with your thumb, leading to greater force and pressure!

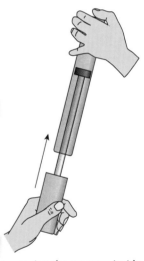

GAS TEMPERATURE AND MEAN KINETIC ENERGY

Gas particles are continually moving at different velocities inside the container. When heat energy is added to the gas, this increases both the **velocity** and the kinetic energy of the particles. The gas temperature T is proportional to the mean kinetic energy of these particles (but not the velocity).

Increasing the pressure inside a bicycle pump

AIR PRESSURE

The air in the atmosphere extends about 120km above us. The pressure exerted by gas particles in the air is known as atmospheric pressure. Millions of gas particles are continually colliding with the Earth's surface (and us!), each one exerting a tiny force. The result of all of these collisons is to produce atmospheric (or air) pressure. The value for normal atmospheric pressure is 101 000 Pa (1.01×10^5 Pa) . This is equivalent to a force of 100 000 N acting on one square metre of surface.

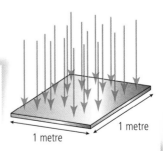

Air particles colliding with a surface

GAS LAWS AND KINETIC MODEL 2

This section follows on from the work on pressure. The key concepts to learn in this topic are:

- knowledge of the relationship between Kelvin and degrees Celsius and the absolute zero of temperature:

  $0K = -273°C$

- explanation of the pressure–volume, pressure–temperature and volume–temperature laws qualitatively in terms of a kinetic model

- use of appropriate relationships to solve problems involving the volume, pressure and temperature of a fixed mass of gas:

  $p_1 V_1 = p_2 V_2$, $\frac{p_1}{T_1} = \frac{p_2}{T_2}$, $\frac{V_1}{T_1} = \frac{V_2}{T_2}$, $\frac{pV}{T} = constant$

Gas particles in a sealed container

- description of experiments to verify the pressure–volume law (Boyle's law), the pressure–temperature law (Gay-Lussac's law) and the volume–temperature law (Charles' law).

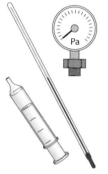

Measuring temperature, pressure and volume

VIDEO LINK

For another example of temperature affecting air pressure, watch the experiment with a hardboiled egg at www.brightredbooks.net/N5Physics

DON'T FORGET

Each experiment is always carried out using a fixed mass.

DESCRIPTION OF EXPERIMENTS TO VERIFY THE THREE GAS LAWS

When studying the properties of gases in containers, the temperature (T), pressure (p) and volume (V) can be easily measured using thermometers, pressure gauges and a gas syringe.

Measuring properties of gases

A gas in a sealed container has a *fixed* mass. So the number of gas particles (whether molecules or atoms) is fixed. There are three properties of the gas in the container:

- pressure, p
- temperature, T
- volume, V

When an inflated balloon is moved from a cold to a hotter spot its volume increases

A pressure cooker has a release valve to let hot steam out to reduce the pressure

Three separate experiments can be carried out to determine the relationship between p, T and V. In each experiment, one variable is kept constant to find how the remaining two variables are related:

Known as:

- Experiment 1: V is constant; p and T are variable — Gay-Lussac's Law (Pressure Law)
- Experiment 2: p is constant; V and T are variable — Charles' Law
- Experiment 3: T is constant; p and V are variable — Boyle's Law

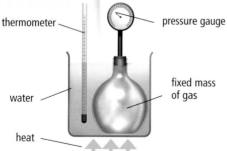

thermometer

pressure gauge

water

heat

fixed mass of gas

Measuring the effect of temperature on gas pressure at constant volume

DON'T FORGET

The mass and volume of the gas are constant.

EXPERIMENT 1: PRESSURE AND TEMPERATURE (GAY-LUSSAC'S LAW)

Here, we're investigating the relationship between pressure (p) and temperature (T) for a fixed mass of gas at *constant volume*.

In this experiment the temperature of gas inside a circular flask is changed and the changing gas pressure is recorded. It is assumed that the volume of the glass flask does not change when heated, and so the volume of the gas inside is also constant.

When a graph of pressure versus temperature is drawn, a straight line is obtained, but does not pass through the origin.

contd

Changing temperature and pressure at constant volume

The straight line on the graph has been projected until it reaches the temperature axis. Zero pressure is at –273°C. Zero pressure indicates the *true zero* of temperature. A lower temperature than this is not possible, which is why –273°C is known as **absolute zero**.

When the pressure versus temperature graph is redrawn with the temperature starting at –273°C, an alternative temperature scale called the *Kelvin scale* can be defined in which absolute zero is given as 0K.

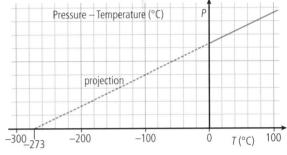

Pressure–temperature (°C) graph

Absolute zero of temperature

- The Kelvin scale uses the same divisions as the Celsius scale.
- So, one degree Celsius is the same size as one kelvin.
- The actual zero of the Kelvin scale is –273·15°C.

Converting between temperature scales

For most calculations the approximate value –273°C is used.

- To convert degrees Celsius into kelvin just add 273. For example, 25°C = 25 + 273 = 298K (the kelvin unit is simply K – no degrees!)
- To convert kelvin into degrees Celsius subtract 273. For example, 300K = 300 – 273 =27°C.

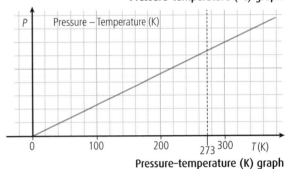

Pressure–temperature (K) graph

For Experiment 1, on the Kelvin scale, the straight line graph through the origin means that pressure is directly proportional to temperature in kelvin:

$$p \propto T \quad \frac{p}{T} = k \text{ (where k = constant)} \quad \frac{p_1}{T_1} = \frac{p_2}{T_2} \quad \text{at constant volume.}$$

This is the Gay-Lussac's Law (or the *Pressure Law*).

Whenever this equation is used, the Celsius temperature of the gas must be converted into kelvin.

EXAMPLE:

A sample of gas in a sealed container is at a pressure of 200 kPa and at a temperature of 135°C. Calculate the new pressure if its temperature is reduced to 78°C and the volume of the container remains constant.

$p_1 = 200\,kPa, p_2 = ?, T_1 = 135°C, T_2 = 78°C$

Convert T_1 and T_2 into kelvin for use in the equation:

$T_1 = 135°C = 135 + 273 = 408K, T_2 = 78°C = 78 + 273 = 351K$

$\frac{p_1}{T_1} = \frac{p_2}{T_2} \Rightarrow \frac{200}{408} = \frac{p_2}{351}$

$p_2 \times 408 = 200 \times 351$

$p_2 = 172\,kPa$

Note that in this example, the pressure can remain in kPa in the calculation, but the temperature must always be converted from Celsius into kelvin.

 ## THINGS TO DO AND THINK ABOUT

1. Use the Pressure Law to explain the following examples:

- Car tyre pressures need to be checked when winter weather arrives.
- When tackling industrial blazes, firefighters must be extremely careful – especially when there are gas cylinders present inside the factory.

GAS LAWS AND THE KINETIC MODEL 3

DON'T FORGET

The mass and pressure of the gas are constant.

EXPERIMENT 2: VOLUME AND TEMPERATURE (CHARLES' LAW)

Here we're investigating the relationship between volume (V) and temperature (T) for a fixed mass of gas at constant *pressure*.

In this experiment the temperature of gas trapped in a long, narrow column of glass is altered and the changing volume of the gas is recorded. The water is heated slowly, so the gas inside the glass is the same temperature as the water.

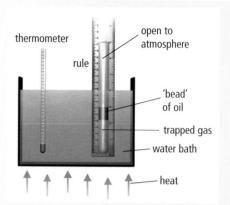

Investigating the effect of temperature on the volume of a fixed mass of gas at constant pressure

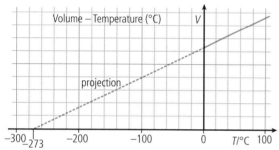

Volume–temperature (°C) graph

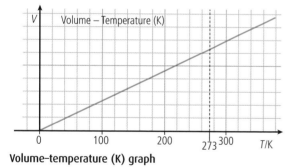

Volume–temperature (K) graph

The bead inside the glass column moves so that the pressure of the trapped gas stays in equilibrium with atmospheric pressure (so pressure remains constant). When a graph of volume versus temperature is drawn, a straight line is obtained, but does not pass through the origin.

The straight line on the graph has been projected until it reaches the temperature axis. Zero volume appears at −273 °C. As we have seen, zero pressure indicates the true zero of temperature.

When the graph is redrawn with the temperature starting at −273°C, a new scale can be defined where absolute zero is 0K – the Kelvin scale.

On this Kelvin scale, the straight line graph through the origin means that volume is directly proportional to temperature in kelvin:

$$V \propto T \qquad \frac{V}{T} = k \text{ (where } k = \text{constant)} \qquad \frac{V_1}{T_1} = \frac{V_2}{T_2} \text{ at constant pressure.}$$

This is known as *Charles' law*.

Whenever this equation is used, Celsius temperatures must be converted into kelvin.

VIDEO LINK

For another explanation of Charles' Law, watch the video at www.brightredbooks.net/N5Physics

EXAMPLE 1

A party balloon is inflated to a volume of 336 cm³ and sealed at a temperature of 29°C. During the night, its temperature falls to 3°C.

(a) Calculate the new volume of the balloon.

$V_1 = 336cm^3$, $V_2 = ?$, $T_1 = 29°C$, $T_2 = 3°C$

Convert T_1 and T_2 into kelvin for use in the equation:

$T_1 = 29°C = 29 + 273 = 302K$, $T_2 = 3°C = 3 + 273 = 276K$

$$\frac{V_1}{T_1} = \frac{V_2}{T_2} \qquad \frac{336}{302} = \frac{V_2}{276}$$

$V_2 \times 302 = 336 \times 276 = 307 \cdot 1cm^3$

Note that in this question, the volume can remain in cm³ in the calculation, but the temperature must be converted from Celsius into kelvin.

(b) State any assumptions that you have made.

The answer to part (a) assumes that no air leaked from the balloon and that the air pressure remained constant overnight while the balloon reduced in size.

EXPERIMENT 3: VOLUME AND PRESSURE (BOYLE'S LAW)

Here we're investigating the relationship between volume (V) and pressure (p) for a fixed mass of gas at constant temperature. In this experiment the pressure of gas inside a syringe is changed by pressing the plunger and each new volume of the gas is recorded. The plunger is pressed slowly, so the gas inside the syringe stays at constant temperature.

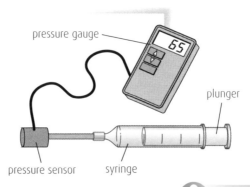

pressure gauge

plunger

pressure sensor   syringe

When a graph of volume versus pressure is drawn, a curve is obtained, which shows that as the pressure increases, the volume decreases.

When a graph of 1/volume versus pressure is drawn, a straight line through the origin is obtained.

This means that the pressure is inversely proportional to the volume:

$$p \propto \frac{1}{V} \qquad pV = k \text{ (where k = constant)}$$

$$p_1V_1 = p_2V_2$$

at constant temperature.

This is known as *Boyle's law*.

DON'T FORGET

The mass and temperature of the gas are constant.

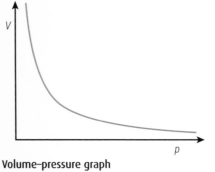

Volume–pressure graph

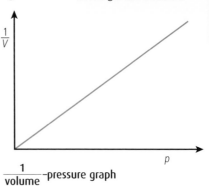

$\dfrac{1}{\text{volume}}$–pressure graph

EXAMPLE 2

Gas contained in a syringe has a volume of $55\,cm^3$ and is at a pressure of $2.34 \times 10^4\,Pa$. The syringe is compressed until the new volume is $27\,cm^3$. Assume that the temperature remains constant.

(a) Calculate the new pressure of the gas in the syringe.

$V_1 = 55cm^3$, $V_2 = 27cm^3$, $p_1 = 2.34 \times 10^4$ Pa, $p_2 = ?$

$p_1V_1 = p_2V_2$

$2.34 \times 10^4 \times 55 = p_2 \times 27$

$p_2 = \dfrac{2.34 \times 10^4 \times 55}{27} = 4.77 \times 10^4$ Pa

(b) In practice, as it is compressed, the temperature of the gas rises. Explain what effect this may have on the final pressure of the gas.

If the gas temperature rises, then the kinetic energy and **average velocity** of the gas particles will increase. This will cause more collisions and greater force of collisions, causing the pressure to increase.

ONLINE

Visit www.brightredbooks.net/N5Physics to test your knowledge of the relationship between p, T and V.

VIDEO LINK

Watch the unusual properties of helium when 'supercooled' at www.brightredbooks.net/N5Physics

 THINGS TO DO AND THINK ABOUT

1. Consider which individual gas equation explains each of these phenomena:
   - A football, inflated indoors and then taken outside on a cold day, shrinks slightly.
   - Deep sea divers breathe a mixture of oxygen and helium when submerged, and sometimes return to the surface in stages.
   - Deep sea fish die when they are brought to the surface, even if kept in water tanks.
   - People's ears sometimes 'pop' in an aircraft at take-off and landing.
   - A hot air balloon rises when the trapped air inside the balloon is heated.

2. Research the uses of 'superconductors' – when wires conduct at extremely low temperatures.

GAS LAWS AND THE KINETIC MODEL 4

DON'T FORGET

Remember to convert temperatures into kelvin!

DON'T FORGET

The kinetic theory of gases is when the behaviour of the particles in a gas is used to explain the relationship between the pressure, volume and temperature of the gas.

VIDEO LINK

Watch the simulation of gas particles at www.brightredbooks.net/N5Physics

THE GENERAL GAS EQUATION

The three gas laws can be combined into one *general gas equation*, in which p, V and T are all variable.

1: $\dfrac{p_1}{T_1} = \dfrac{p_2}{T_2}$ 2: $\dfrac{V_1}{T_1} = \dfrac{V_2}{T_2}$ 3: $p_1 V_1 = p_2 V_2$

The three equations above combine to make the general gas equation:

$\dfrac{pV}{T} = k$ (where k = constant)   or $\dfrac{p_1 V_1}{T_1} = \dfrac{p_2 V_2}{T_2}$

EXAMPLE 1

A weather balloon is inflated to $2.83\,m^3$ at the Earth's surface at a pressure of $1.01 \times 10^5\,Pa$. The temperature of the air inside the balloon is 28°C. When the balloon is released into the atmosphere it rises to an altitude of 47 km where the air pressure is 111 Pa and the volume of the balloon increases to $269\,m^3$.

Calculate the air temperature at this altitude.

$V_1 = 2.83\,m^3$, $V_2 = 269\,m^3$, $T_1 = 28°C$, $T_2 = ?$, $p_1 = 1.01 \times 10^5$ Pa, $p_2 = 111$ Pa

Convert T_1 into kelvin for use in the equation: $T_1 = 28°C = 28 + 273 = 301K$

$\dfrac{p_1 V_1}{T_1} = \dfrac{p_2 V_2}{T_2}$

$\dfrac{1.01 \times 10^5 \times 2.83}{301} = \dfrac{111 \times 269}{T_2}$ so $T_2 = \dfrac{111 \times 269 \times 301}{1.01 \times 10^5 \times 2.83} = 31.4$ K

EXAMPLE 2

Car engines contain cylinders filled with air. As part of the engine's combustion process, a piston is used to compress the air inside the cylinder, causing an increase in its temperature.

During a test of one compression of air inside the cylinder, the following results were obtained:

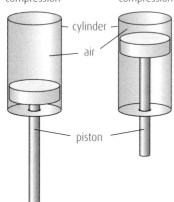

before compression

after compression

cylinder

air

piston

	before compression	after compression
pressure of air in cylinder (Pa)	1.0×10^5	1.2×10^7
volume of air in cylinder (m^3)	4.0×10^{-4}	
temperature of air in cylinder (°C)	18.0	330

Calculate the final volume of the air inside the cylinder.

First, convert both temperatures into Kelvins by adding 273:

$T_1 = 18 + 273 = 291$ K ; $T_2 = 330 + 273 = 603$ K

$\dfrac{p_1 \times V_1}{T_1} = \dfrac{p_2 \times V_2}{T_2}$

$\dfrac{1.0 \times 10^5 \times 4.0 \times 10^{-4}}{291} = \dfrac{1.2 \times 10^7 \times V_2}{603}$

so

$V_2 = \dfrac{603 \times 1.0 \times 10^5 \times 4.0 \times 10^{-4}}{1.2 \times 10^7 \times 291} = 6.9 \times 10^{-6}$ m^3

THE GAS LAWS AND THE KINETIC MODEL

Each of the gas laws can be explained in terms of the behaviour of gas particles.

1 The pressure-temperature law (Guy-Lussac's law, constant volume, p ∝ T)

If the gas temperature increases, the particles have more kinetic energy and they move faster. The particles hit the walls with greater average force. Also, because the volume is constant, they hit the container walls more often. These two effects cause an increase in pressure.

2 The volume-temperature law (Charles' law, constant pressure, V ∝ T)

If the temperature increases, the particles have more kinetic energy and they move faster. The particles hit the container walls with more force which causes the gas pressure to increase. This increase in pressure causes the gas volume to increase, which results in fewer collisions each second. When the inside pressure is the same as the outside pressure the volume stops increasing.

The pressure-volume law (Boyle's Law, constant temperature)

As the temperature is constant, the average kinetic energy and the velocity are constant. This means that the particles hit the walls of the container with the same average force. When the gas volume is increased, the particles have to travel further between collisions with the container walls. There are fewer particle collisions per second. This causes the pressure to fall.

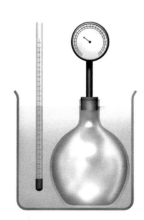

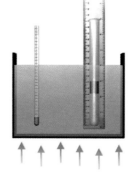

Changing temperature and pressure of a constant volume

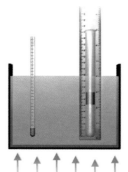

Changing temperature and volume at constant pressure

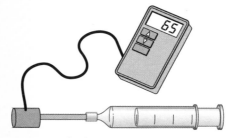

Changing pressure and volume at constant temperature

EXAMPLE 3

The relationship between the pressure and temperature of a fixed mass of gas at constant volume is investigated using the apparatus shown.

The temperature and pressure of the gas increase as the gas is slowly heated.

Use the kinetic model to explain change in gas pressure as the temperature increases.

As temperature increases, the average E_k of the gas particles increases, so the particles collide with the walls of the container more frequently and with greater force.

Since $p = \dfrac{F}{A}$, this results in an increased pressure.

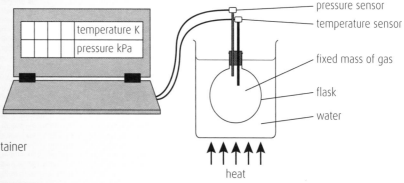

temperature K
pressure kPa

pressure sensor
temperature sensor
fixed mass of gas
flask
water
heat

THINGS TO DO AND THINK ABOUT

1. Atmospheric weather balloons rise to enormous altitudes, to capture data required for weather forecasting. The air pressure and temperature reduce at such altitudes. Carry out research to find out the balloon dimensions, what data they collect and what eventually happens to them.

WAVES

WAVE PARAMETERS AND BEHAVIOURS 1

The key concepts to learn in this topic are:
- knowledge that waves transfer energy
- definition of transverse and longitudinal waves
- knowledge that sound is an example of a longitudinal wave and electromagnetic radiation and water waves are examples of transverse waves

- determination of the frequency, period, wavelength, amplitude and wave speed for longitudinal and transverse waves
- the use of appropriate relationships to solve problems involving wave speed, frequency, period, wavelength, distance, number of waves and time:

$$v = \frac{d}{t}, \quad f = \frac{N}{t}, \quad v = f\lambda, \quad T = \frac{1}{f}$$

ONLINE

For more details on the key concepts for this topic, go to www.brightredbooks.net/N5Physics

DON'T FORGET

Transverse waves: particles vibrate at right angles to the wave's direction of travel. Longitudinal waves: particles vibrate in the same direction as the wave's direction of travel.

a = amplitude λ = wavelength

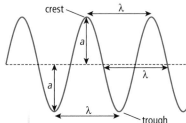

The diagram shows how the various parts of a wave are usually represented

VIDEO LINK

Check out the clips on particle motion at www.brightredbooks.net/N5Physics

DON'T FORGET

Energy moves through a medium; the particles just vibrate.

DON'T FORGET

You need to be able to work out amplitude and wavelength from the type of wave diagram in Example 1.

THE TRANSFER OF ENERGY IN WAVES

Waves are created by **vibration**. Waves transfer energy from one place to another. For example, when waves reach the seashore, the water particles are vibrating. (This means that the water particles have kinetic energy (E_k) when they are moving and potential energy (E_p) when they are displaced from their resting position.)

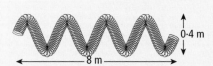

Waves crashing ashore can transfer so much energy that they cause damage

TYPES OF WAVES

There are two types of waves:

1. Longitudinal waves

In longitudinal waves, the particles of the medium vibrate in the same direction as the wave travels.

Sound waves are an example of longitudinal waves.

2. Transverse waves

In transverse waves, the particles of the medium vibrate at right angles to the direction of wave travel.

Electromagnetic radiation and water waves are examples of transverse waves.

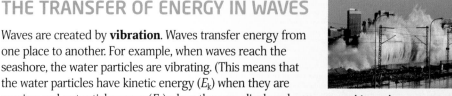

longitudinal wave — wave direction →
← → particle direction

transverse wave

particle direction   wave direction →

In longitudinal waves each spring particle moves back and forth around a central position

In transverse waves the spring particles move up and down around a central position

These diagrams show the direction of wave travel and the direction of vibration of the particles

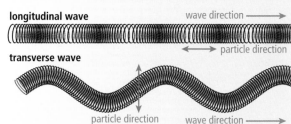

water waves viewed from the side

water waves viewed from above

Long wavelength   Short wavelength

Water waves, when viewed from above, can be represented by a series of lines that are straight or curved depending on the wave pattern. Each line represents the **crest** of a wave. The distance between two lines is one wavelength.

EXAMPLE 1

A slinky is used to show some wave properties. A simplified diagram of a wave produced by a slinky is shown here.

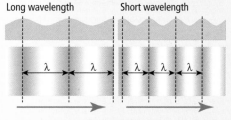

0·4 m

8 m

(a) Determine the amplitude of the wave. $a = \frac{0·4}{2} = 0·2\,m$

(b) Calculate the wavelength of the wave.

There are four complete waves in 8 m. So, d = length all four waves = 8 m.

N = number of complete waves = 4 waves

$\lambda = \frac{d}{N} = \frac{8}{4} = 2\,m$

contd

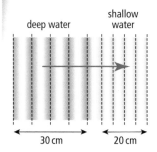

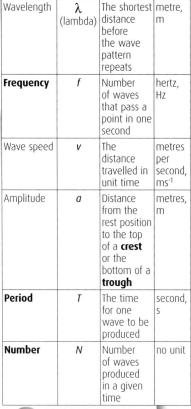

EXAMPLE 2

A ripple tank is a device that is used to demonstrate the properties of water waves. A ripple tank was used to show water waves travelling from deep to shallow water. The diagram represents the water waves before and after they reach the shallow water.

Calculate the wavelength of the water waves: (a) in deep water (b) in shallow water.

(a) One wave between each crest → 4 waves in 30 cm   Deep: $\lambda = \dfrac{d}{N} = \dfrac{30}{4} = 7 \cdot 5\,cm$

(b) One wave between each crest → 4 waves in 20 cm   Shallow: $\lambda = \dfrac{d}{N} = \dfrac{20}{4} = 5 \cdot 0\,cm$

Water waves

EXAMPLE 3

Water waves are observed travelling in the sea. The crest of one wave travelled 150 m in 45 seconds.

Determine the speed of the waves.

$$\text{wave speed } V = \frac{d}{t} = \frac{150}{45} = 3 \cdot 3\ ms^{-1}$$

DON'T FORGET

Greater wave amplitude = greater energy transferred.

ENERGY OF A WAVE

The energy transferred by waves depends on their amplitude.

The greater the wave amplitude, the greater the energy transferred

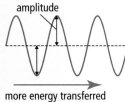

amplitude

more energy transferred

amplitude

less energy transferred

Quantity	Symbol	Definition	Unit
Wavelength	λ (lambda)	The shortest distance before the wave pattern repeats	metre, m
Frequency	f	Number of waves that pass a point in one second	hertz, Hz
Wave speed	v	The distance travelled in unit time	metres per second, ms^{-1}
Amplitude	a	Distance from the rest position to the top of a **crest** or the bottom of a **trough**	metres, m
Period	T	The time for one wave to be produced	second, s
Number	N	Number of waves produced in a given time	no unit

WAVE DEFINITIONS

Wave frequency, f, is calculated by counting the number of waves produced in a certain time, then calculating the number produced each second.

The period, T, and frequency, f, of waves are connected by the relationship: $f = \dfrac{1}{T}$

Using this equation, frequency can be calculated if the period of the wave is known.

EXAMPLE 4

A tuning fork produces 41 880 vibrations in 2 minutes. Calculate the frequency of the note produced.

frequency = number of vibrations each second: $f = \dfrac{\text{number of vibrations}}{\text{time taken in seconds}} = \dfrac{41880}{2 \times 60} = 349\,Hz$

EXAMPLE 5

The **oscilloscope** wave pattern represents an electrical signal.

Determine: (a) the amplitude of the signal

amplitude, a = distance from centre to top of signal
$\qquad$ = 30 mV

(b) the frequency of the signal.

period, T, of wave from graph = 6 μs

$$f = \frac{1}{T} = \frac{1}{6 \times 10^{-6}} = 1 \cdot 7 \times 10^{5}\ Hz$$

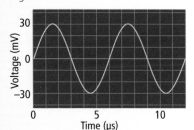

DON'T FORGET

When using frequency and time, always covert into hertz and seconds.

ONLINE TEST

Test yourself at www.brightredbooks.net/N5Physics

THINGS TO DO AND THINK ABOUT

1. Research how wave energy can be harnessed by offshore wave generators.
2. State the advantages and disadvantages of installing these devices.

WAVE PARAMETERS AND BEHAVIOURS 2

The key concepts to learn in this topic are:
- use of appropriate relationships to solve problems involving wave speed, frequency, period, wavelength, distance, number of waves and time:

$$v = \frac{d}{t}, \; f = \frac{N}{t}, \; v = f\lambda, \; T = \frac{1}{f}$$

- knowledge that diffraction occurs when waves pass through a gap or around an object
- comparison of long wave and short wave diffraction
- drawing diagrams using wavefronts to show diffraction when waves pass through a gap or around an object.

THE WAVE EQUATION

Wavespeed, v, is the distance travelled by a wave in one second. To calculate **wavespeed** you can use one of two methods.

Method 1

The wavespeed, v, can be calculated by multiplying the wavelength, λ, by the frequency, f.

$$v = f\lambda$$

This is know as the **wave equation**.

Method 2

The wavespeed, v, can also be calculated by dividing the distance travelled by the wave, d, by the time, t.

$$v = \frac{d}{t}$$

EXAMPLE 1

A person aboard a boat anchored at sea counts 24 waves passing the boat in 2 minutes. The time for a wave crest to travel from the boat to a buoy is 42 s. The buoy is 140 m from the boat.

Determine:
(a) the speed of the waves

The waves travel 140 m in 42 seconds.

$$v = \frac{d}{t} = \frac{140}{42} = 3\cdot3 \; \text{m s}^{-1}$$

(b) the frequency of the waves

24 waves pass the buoy in 120 seconds.

$$f = \frac{\text{number of waves}}{\text{time taken}} = \frac{24}{120} = 0\cdot2 \, \text{Hz}$$

(c) the wavelength of the waves.

$$\lambda = \frac{v}{f} = \frac{3\cdot3}{0\cdot2} = 16\cdot5 \, \text{m}$$

EXAMPLE 2

The 'sodium doublet' is the name given to two spectral lines of the element sodium which are used by optical scientists as 'benchmarks' when checking instruments that measure wavelength. The doublet consists of two yellow lines close together in the **visible spectrum**.

The wavelengths of the lines are 589·0 nm and 589·6 nm. Calculate the difference in frequencies of these lines.

Both waves travel at speed of light in a vacuum = 3×10^8 m s^{-1}

$$f_1 = \frac{v}{\lambda} = \frac{3 \times 10^8}{589 \times 10^{-9}} = 5\cdot093 \times 10^{14} \, \text{Hz}$$

$$f_2 = \frac{v}{\lambda} = \frac{3 \times 10^8}{589\cdot6 \times 10^{-9}} = 5\cdot088 \times 10^{14} \, \text{Hz}$$

$$\Delta f = f_1 - f_2 = 5\cdot093 \times 10^{14} - 5\cdot088 \times 10^{14} = 5\cdot0 \times 10^{11} \, \text{Hz}$$

contd

DIFFRACTION

All waves display several common behaviours:

- They all transfer energy.
- They all exhibit the properties of reflection, refraction, diffraction and interference.

Diffraction is the property of waves which occurs when they *bend* around obstacles. Diffraction can be demonstrated using water waves.

Diffraction around obstacles

Waves diffract (bend) into the gaps behind obstacles when they pass by. Note that when waves are diffracted, their *wavelength* does not change. But, the longer the wavelength, the greater the diffraction.

A practical example of this concerns television and radio signal reception.

Television and **radio signals** travel as **electromagnetic waves** and both travel at the speed of light. **Television waves** have a much greater frequency than **radio waves**, and so have a much shorter wavelength.

A house situated behind a hill from the **transmitter** (as shown in the diagram) relies on diffraction of the signals around the hilltop to reach the aerial. Radio waves have much longer wavelengths than TV waves, and so diffract more. So the house receives radio signals, but not TV signals.

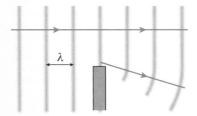
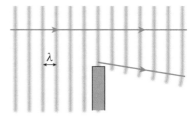

Low-frequency long wavelengths: most diffraction.

High-frequency short wavelengths: least diffraction.

Diffraction around obstacles

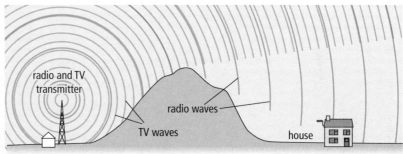

Television and radio reception is affected by the wavelength of the respective signals

Diffraction through gaps

Diffraction also occurs when waves pass through gaps.

When waves pass through a gap, the width of the gap determines how much diffraction occurs:

- If the gap is less than or equal to the wavelength of the waves passing through (gap ⩽ 1 λ), circular waves emerge from the gap.
- If the gap is greater than the wavelength (gap > λ), straight waves emerge, diffracted at their edges.

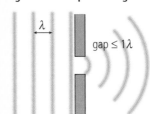

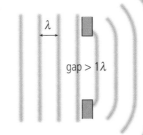

Diffraction through gaps

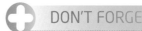

DON'T FORGET

When waves are diffracted, the wavelength does not change.

 VIDEO LINK

Check out the 'Exploring Wave Motion' clip at www.brightredbooks.net/N5Physics

 ONLINE TEST

How well have you learned about the wave equation and wave properties? Go online and test yourself at www.brightredbooks.net/N5Physics

 THINGS TO DO AND THINK ABOUT

1. Investigate some phenomena caused by diffraction. For example:

(a) Why can we hear sounds from behind obstacles, but cannot see light from behind obstacles?

(b) Why does thunder sound like a low rumble when a storm is distant, but sound like a sharp crack when the storm is close by?

ELECTROMAGNETIC SPECTRUM 1

The key concepts to learn in this topic are:

- knowledge of the relative frequency and wavelength of bands of the electromagnetic spectrum
- knowledge of typical sources, detectors and applications for each band in the electromagnetic spectrum
- knowledge that all radiations in the electromagnetic spectrum are transverse and travel at the speed of light.

AN OVERVIEW

The **electromagnetic (EM) spectrum** is the name given to a family of waves. These waves do *not* require any moving particles to transfer their energy. Electromagnetic waves consist of vibrating electric and magnetic fields. The energy transferred by these waves is referred to as electromagnetic *radiation*.

The different parts of the EM spectrum have the following properties in common:

- they are all able to travel through a vacuum
- they all transfer energy
- they all travel at the speed of light in a vacuum ($3 \times 10^8\,\mathrm{m\,s^{-1}}$)
- they all exhibit wave properties of reflection, refraction, diffraction and interference and obey the wave equation ($v = f\lambda$).

Although they all travel at the same speed, different parts of the spectrum have different *wavelengths* and *frequencies*.

This diagram of the EM spectrum displays the names, relative wavelengths and frequencies of its different parts.

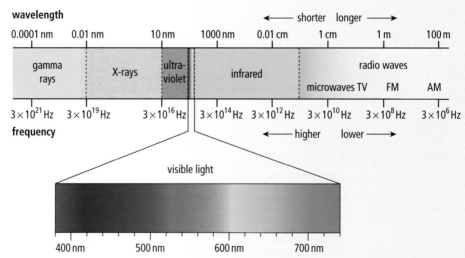

The electromagnetic spectrum

The order of the seven different parts of the spectrum is determined by their wave frequency.

The frequency and wavelength are inversely proportional. In the EM spectrum, as the wave frequency increases, the wavelength decreases.

contd

 DON'T FORGET

All parts of the electromagnetic spectrum travel at the speed of light.

DON'T FORGET

The electromagnetic spectrum consists of transverse waves.

 DON'T FORGET

You should memorise the different parts of the EM spectrum, including which has smallest wavelength and highest frequency.

EXAMPLE 1

UVB ultraviolet radiation can cause tanning, sunburn and skin cancer. UVB wavelengths range from 280–320 nm. Calculate the frequency range of these waves.

All EM spectrum waves travel at 3×10^8 m s^{-1}.

$\lambda_1 = 280 \times 10^{-9}$ m

$f_1 = \dfrac{v}{\lambda}$

$\quad = \dfrac{3 \times 10^8}{280 \times 10^{-9}}$

$\quad = 1 \cdot 07 \times 10^{15}$ Hz

$\lambda_2 = 320 \times 10^{-9}$ m

$f_2 = \dfrac{v}{\lambda}$

$\quad = \dfrac{3 \times 10^8}{320 \times 10^{-9}}$

$\quad = 9 \cdot 38 \times 10^{14}$ Hz

So the frequency range is $9 \cdot 38 \times 10^{14}$ Hz to $1 \cdot 07 \times 10^{15}$ Hz.

DON'T FORGET

The greater the frequency, the greater the energy of the waves in the electromagnetic spectrum.

EXAMPLE 2

The electromagnetic spectrum is shown here in order with some radiations missing.

gamma	A	ultraviolet	B	infrared	microwaves	radio waves

(a) Name radiations A and B.

A = X-rays, B = visible light.

(b) Which radiation has the highest frequency?

Gamma radiation has the highest frequency.

ONLINE TEST

Take the 'Electromagnetic spectrum' test online at www.brightredbooks.net/ N5Physics

EXAMPLE 3

The frequencies of some colours of light are shown in the table.

Colour	frequency (Hz)
blue	$7 \cdot 5 \times 10^{14}$
yellow	$6 \cdot 0 \times 10^{14}$
red	$4 \cdot 5 \times 10^{14}$

(a) State the speed of light in air.

$3 \cdot 0 \times 10^8$ ms^{-1}

(b) Show by calculation which colour has a wavelength of 5.0×10^{-7} m.

$v = f\lambda, \lambda = \dfrac{v}{f} = \dfrac{3 \cdot 0 \times 10^8}{5 \cdot 0 \times 10^{-7}} = 6 \cdot 0 \times 10^{14}$,

so colour is yellow.

ONLINE

For more information, follow the electromagnetic spectrum link at www.brightredbooks.net/ N5Physics

EXAMPLE 4

Microwaves are transmitted from an orbiting satellite to an Earth ground station.

The microwaves have a frequency of $6 \cdot 0 \times 10^9$ Hz.

(a) Determine the period of the microwaves.

$T = \dfrac{1}{f} = \dfrac{1}{5 \cdot 0 \times 10^9} = 2 \cdot 0 \times 10^{-10}$ s

(b) Determine the wavelength of the microwaves.

$v = f\lambda, \lambda = \dfrac{v}{f} = \dfrac{3 \times 10^8}{6 \cdot 0 \times 10^9} = 0 \cdot 05$ m

(c) The satellite also sends radio waves to the Earth station at the same time.

Will the radio waves be received after, before or at the same time as the microwaves?

The waves will be received at the same time.

Radio waves and microwaves travel at the same speed through space.

THINGS TO DO AND THINK ABOUT

The Sun and stars all give out radiation from the EM spectrum, although the Earth's atmosphere filters out some parts of the spectrum before they reach the surface.

1. Find out which parts of the atmosphere absorb which parts of the spectrum.

2. How is research carried out on the radiations which do not reach the surface of the Earth?

ELECTROMAGNETIC SPECTRUM 2

TYPICAL SOURCES, DETECTORS AND APPLICATIONS FOR EACH BAND OF THE ELECTROMAGNETIC RADIATION

The different radiations of the electromagnetic spectrum have many applications. The Sun and stars are sources of all radiations present in the electromagnetic spectrum. However, the most common applications involve sources which are produced on Earth.

	Typical sources	Detector	Applications and uses	Additional facts
Gamma rays	· Radioactive substances · Nuclear reactors	· **Geiger–Muller tube** and **counter**	· Medical diagnosis using **radioisotopes** as **gamma ray** tracers · Sterilisation of medical instruments · Scanning shipping containers at ports	· Highest energy of EM spectrum · Cannot penetrate Earth's atmosphere from outer space · Causes **ionisation**, which can affect living tissue
X-rays	· **X-rays** are caused when very fast moving electrons collide with a metal target · X-ray machines	· X-rays darken photographic film · X-ray image intensifiers · Geiger–Muller tube and counter	· Medical imaging, giving 3D images of the internal body structure, used especially in the diagnosis of broken bones · Analysis of atomic structures using X-ray crystallography · Airport security scanners	· High energy waves · Cannot penetrate Earth's atmosphere from outer space · Causes ionisation, which can affect living tissue
Ultraviolet radiation	· The Sun is an important source of **UV radiation** · UV lamps, including mercury vapour lamps	· UV radiation causes some materials to 'fluoresce' i.e. glow when exposed to UV · A UV photodiode is an electronic device which detects UV	· UV radiation causes a chemical reaction in the skin that produces the important nutrient vitamin D · UV radiation is used in the treatment of certain skin conditions · Disinfection of hospital equipment · Used by dentists to 'cure' or harden composite material used for fillings	· Most UV radiation from the Sun is absorbed in the upper atmosphere by the ozone layer · The UV radiation range is sometimes separated into three bands: UVA, UVB and UVC · UVC is the highest frequency UV (hence the highest energy) and, fortunately, is filtered out by the ozone layer · UVA and UVB radiation cause sunburn and tanning, and overexposure can cause skin cancer · Causes ionisation, which can affect living tissue
Visible light	· The Sun is a primary source of visible light · Light bulbs, lasers	· Photodiodes · Phototransistors · Light dependent resistors detect light energy	· Light from lasers is widely used in communication through optical fibres · Supermarket checkout readers use laser light to scan barcodes for information · Laser light is used in medical treatment and surgery	· The eye responds to visible light, which occupies the smallest range of wavelengths in the EM spectrum · Red light has the longest and blue light has the shortest wavelength in the spectrum

contd

	Typical sources	Detector	Applications and uses	Additional facts
Infrared rays	· **Infrared radiation** is received from the Sun · Infrared heaters	· (Black bulb) thermometer · Thermopile	· Used as heating source · Humans emit infrared radiation – 'night/thermal image' IR cameras are used to detect infrared radiation in darkness, and to locate trapped disaster survivors · Used in medical diagnosis and treatment · Passive infrared detectors (PID) are used in intruder alarms	· Infrared rays (heat rays) are responsible for heat transfer by radiation · Over half of the radiation received on Earth from the Sun is infrared radiation · Most of the emitted energy from Earth through the atmosphere back into space is in the form of infrared radiation · The surface temperature of the Earth is regulated by the re-radiating of this radiation – this maintains the average temperature of the planet
Microwaves	· Microwave ovens produce waves (called microwaves) with a wavelength of approximately 3 cm	· Radar detector dishes	· Used extensively for communication, for example in **satellite** phones and television outside broadcasts	
Radio waves	· Radio waves are made by various transmitters	· TV aerials · Radar detector dishes	· Used extensively for communication · Transmitters connected to amplifiers produce radio waves	· The wide range of radio frequencies is used for many different types of communication · Radio wave transmitters can range in size from small enough to fit inside a mobile phone, to around 550 m for a TV **transmitter**

EXAMPLE:

Radiation from different bands of the electromagnetic spectrum continually reaches Earth from space. This radiation is studied by astronomers in their research into the origins of the universe.

Explain why detectors on satellites in orbit around the Earth are required to receive some of this radiation.

The Earth's atmosphere absorbs radiation from some bands of the electromagnetic spectrum which arrives from space. This radiation would be difficult to detect on Earth. Satellites orbit the Earth above the atmosphere and so the on-board detectors can receive and detect this radiation. The satellites then transmit this information back to Earth.

VIDEO LINK

For further information on this topic, watch 'The Electromagnetic Spectrum' at www.brightredbooks.net/N5Physics

DON'T FORGET

Sources, applications and detectors for different parts of the electromagnetic spectrum can be asked about in the exam.

THINGS TO DO AND THINK ABOUT

1. Find out more about sources, detectors and uses of the EM spectrum.

2. Try to investigate in more detail how the detectors actually work.

3. Are there any safety considerations which affect the use of electromagnetic radiation? For example, what are the problems, if any, associated with the use of microwave ovens?

ONLINE TEST

Take the 'Electromagnetic spectrum' test online at www.brightredbooks.net/N5Physics

REFRACTION OF LIGHT

The key concepts to learn in this topic are:
- knowledge that refraction occurs when waves pass from one medium to another
- description of refraction in terms of change of wave speed, change in wavelength and change of direction (where the angle of incidence is greater than 0°), for waves passing into both a more dense and a less dense medium
- identification of the normal, angle of incidence and angle of refraction, in ray diagrams showing refraction.

DON'T FORGET

- The normal is a line (usually dotted) drawn at 90° to the edge where the light enters the glass.
- The angle of incidence is the angle between the incident ray and the normal.
- The angle of refraction is the angle between the refracted ray and the normal.

DON'T FORGET

You should be able to recognise the angles of incidence and refraction in diagrams.

DON'T FORGET

When the angle of incidence is 0°, the direction of the refracted ray is unchanged.

REFRACTION

Refraction occurs when waves pass from one medium into another. For example, when light waves pass from air into glass. When light travels from a *less dense* to a *more dense* material (such as from air into glass), *its wavelength and wave speed both decrease*. If the angle of incidence is greater than 0° then the light will change direction (or refract) towards the normal.

When light travels from a *more dense to a less dense* material (such as from glass into air), *its wavelength and wave speed both increase*. If the angle of incidence is greater than 0° then the light will change direction (or refract) away from the normal.

The frequency of the light remains constant.

The amount of refraction depends on the type of materials used.

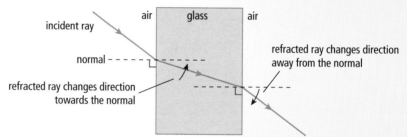

Refraction of light

This diagram illustrates the changes in direction of refracted light

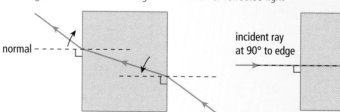

This diagram illustrates that the ray of light would have exactly the same path if it were directed into the glass from the reverse direction.

When the angle of incidence is 0°, the light ray passes straight through

An unusual effect of refraction

The refraction of light can cause some unusual effects. For example, when observing a straight stick that is partly submerged in water, it appears bent.

The explanation involves the refraction of light. When light from the submerged part of the stick reaches the surface, it is refracted away from the normal. When this light enters our eyes, the image of the stick appears to be closer to the surface than it should be.

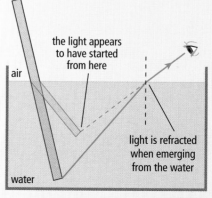

The visual effect of refraction

contd

Lenses

Convex and **concave** lenses are used to refract light for particular applications. Lenses are used in optical instruments, including spectacles, cameras, microscopes and telescopes.

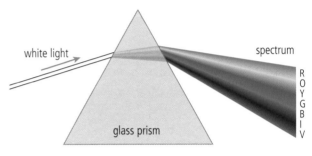

Convex (converging)   Concave (diverging)

Prisms

A **prism** is a three-dimensional triangular piece of glass (or transparent plastic) which is used to refract light.

Different colours of light are refracted by different amounts. A ray of white light directed into a prism will be refracted to show the different colours of the visible spectrum.

Prisms can be used to analyse a light source to determine the colours of light present in the incident ray.

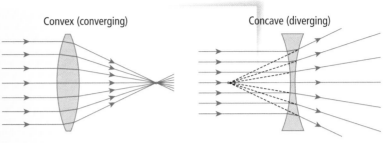

white light   spectrum
glass prism

Light of different colours is refracted by different amounts

REFRACTION EXAMPLES

EXAMPLE 1

Light passes from air into glass. This process is called refraction.

(a) Explain what is meant by refraction.

Refraction is the change of speed and wavelength of waves when they travel from one material into another.

(b) Draw a labelled diagram showing the path of a ray of light being refracted as it passes into a glass block. Label the incident ray, the refracted ray, the normal, the angle of incidence and the angle of refraction.

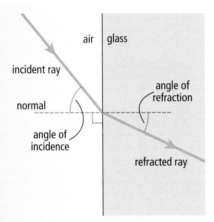

air   glass
incident ray
angle of refraction
normal
angle of incidence
refracted ray

EXAMPLE 2

A ray of green light is incident on a glass block as shown.

Calculate the angle of incidence and angle of refraction.

angle of incidence = 90° − 38° = 52°
angle of refraction = 90° − 59° = 31°

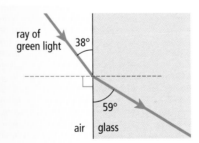

ray of green light   38°
59°
air   glass

VIDEO LINK

Have a look at how refraction works by watching the 'Bill Nye the Science Guy on Light Bending and Bouncing' video at www.brightredbooks.net/N5Physics

ONLINE TEST

Take the 'Light' test online at www.brightredbooks.net/N5Physics

THINGS TO DO AND THINK ABOUT

Refraction of light in glass has many uses in optical instruments, some of which have had a great impact on society.

1. Research the effects on light rays when they pass through convex and concave lenses.

2. Find out how refraction in the glass lenses used in spectacles allows wearers who are short or long sighted to be able to see clearly.

RADIATION

NUCLEAR RADIATION 1

The key concepts to learn in this topic are:
- knowledge of the nature of alpha (α), beta (β) and gamma (γ) radiation
- knowledge of the term 'ionisation' and the effect of ionisation on neutral atoms
- knowledge of the relative ionising effect and penetration of alpha, beta and gamma radiation.

TYPES OF NUCLEAR RADIATION

Some elements are radioactive. This means that they emit radiation. Three types of nuclear radiation are emitted by radioactive elements: alpha, beta and gamma radiation. Some radioactive elements occur naturally and some are created artificially.

An element may have different **isotopes**. Isotopes of an element have different numbers of neutrons in their **nucleus**, but are still the same element. Some elements have isotopes which are radioactive while other isotopes of the element are stable (that is *not* radioactive).

A **radionuclide** is an isotope of an element which is radioactive.

ALPHA (α), BETA (β) AND GAMMA (γ) RADIATION

The three types of nuclear radiation are shown in the diagrams below.

The atoms of radioactive substances emit radiation. Nuclear radiation is emitted from the nucleus at the centre of the atom. All types of radiation can be harmful, but may also be very useful when carefully employed; so it is important to know exactly how each type of radiation behaves.

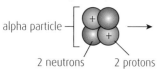

Alpha (α) radiation

alpha particle

2 neutrons   2 protons

Alpha facts

- Alpha particles are large positive nuclei (equivalent to the nuclei of helium atoms).
- They are emitted from radioactive atoms when they disintegrate or decay (give out radiation).
- Alpha particles are slow moving (at approximately 5% of the speed of light).
- They cause greatest ionisation (or greatest ionisation density) because of their large size.

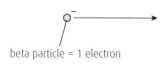

Beta (β) radiation

beta particle = 1 electron

Beta facts

- Beta particles are emitted when neutrons in the nuclei of radioactive atoms are converted into electrons (beta particles) and protons.
- A beta particle is a very fast moving electron (travelling at approximately 90% of the speed of light).
- They cause weak ionisation.

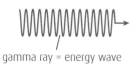

Gamma (γ) radiation

gamma ray = energy wave

Types of radiation

Gamma facts

- Gamma rays are emitted when a radioactive atom disintegrates or decays, sometimes when alpha or beta particles are also emitted.
- Gamma rays are waves of energy and are part of the electromagnetic spectrum.
- They travel at the speed of light.
- Gamma rays are electromagnetic waves and cause very weak ionisation.

IONISATION

Atoms usually have the same number of protons and electrons, so an atom has no overall charge – it is neutral. When alpha, beta or gamma radiation passes through a material, this may cause electrons to be removed or added to some atoms or molecules of the material.

This process is known as **ionisation**.

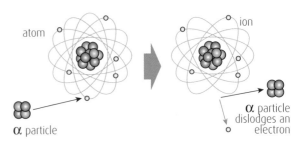

Ionisation of an atom caused by alpha radiation

The atoms or molecules of the material which are affected by the radiation, and the electrons released, become 'ions'. The energy of the radiation is transferred to the material.

- Alpha radiation causes the most ionisation. This happens when an alpha particle collides with an atom removing an electron. The atom then becomes an ion (a charged particle).

- Beta particles cause less ionisation than alpha as they are smaller particles.

- Gamma radiation causes even less ionisation but, because it is a wave of energy, it can travel through atoms without being absorbed.

Unexpected ionisation can be dangerous, and should be prevented. However, controlled use of ionising radiation can be beneficial:

- in the diagnosis and treatment of certain illnesses
- for protecting people, for example by its use in smoke detectors
- in industry, for example to monitor the thickness of newspaper in a paper mill.

VIDEO LINK

Check out the 'Harmful Effects of Radiation' clip at www.brightredbooks.net/N5Physics

DON'T FORGET

Ionisation is when nuclear radiation changes atoms into ions.

PENETRATION OF ALPHA, BETA AND GAMMA RADIATION

DON'T FORGET

Although gamma radiation travels the furthest and is the most penetrating radiation, alpha radiation does the most damage over a very short distance because of its strong ionising capability.

It is vital to know how far radiation can travel (or penetrate) before it is absorbed and is no longer dangerous. This knowledge allows us to select materials for shielding humans from the radiation.

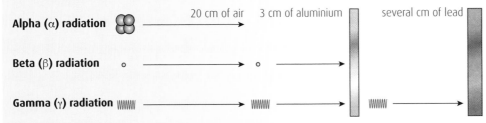

Different types of radiation penetrate materials to different degrees

Alpha radiation can only travel a few centimetres in air before being absorbed. Beta and gamma radiation can travel further through air. A thin sheet of paper or 20 cm of air can absorb alpha radiation. Beta radiation can be absorbed by 3 or 4 cm of aluminium, but gamma requires several centimetres of lead to absorb most of its energy.

THINGS TO DO AND THINK ABOUT

1. Ionisation caused by radiation can be helpful, but may also be dangerous. Carry out research to find particular examples where ionisation has caused major problems and what impact these have had on the people affected by it (you could research the Chernobyl nuclear reactor disaster, for example).

2. Research how a radioactive material which emits alpha radiation is used in some domestic smoke detectors. Explain why it is safe for use in these detectors.

ONLINE TEST

Take the 'Nuclear Radiation' test at www.brightredbooks.net/N5Physics

NUCLEAR RADIATION 2

The key concepts to learn in this topic are:
- definition of activity in terms of the number of nuclear disintegrations and time
- use of an appropriate relationship to solve problems involving activity, number of nuclear disintegrations and time: $A = \dfrac{N}{t}$
- knowledge of sources of background radiation
- knowledge of the dangers of ionising radiation to living cells and of the need to measure exposure to radiation.

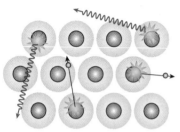

Radioactive decay

ACTIVITY

When a radioactive substance emits radiation, the nuclei of the atoms decay.

The rate of decay is called the **activity** of the substance.

$$\text{Activity} = \frac{\text{number of nuclei decaying}}{\text{time}}$$

$$A = \frac{N}{t}$$

So, the activity of a radioactive source is the number of decays per second.

Activity is measured in **becquerels** (Bq). One becquerel is one decay per second.

EXAMPLE 1

The activity of 1g of uranium is 4·22MBq. How many nuclear disintegrations will occur in 1 minute?

$A = \dfrac{N}{t}$ so $4 \cdot 22 \times 10^6 = \dfrac{N}{60}$ (converting 1 minute into seconds)

$N = 2 \cdot 53 \times 10^8$ decays

ONLINE

For more details on the key concepts of this topic, go to www.brightredbooks.net/N5Physics

DON'T FORGET

Activity is the number of atomic decays every second, expressed in becquerels (bq).

RADIATION DETECTORS

Geiger–Müller tubes

A Geiger–Müller tube can be used to detect radiation. It is usually connected to a counter that counts the number of atoms that have decayed. Each time an atom decays, it is detected by the Geiger-Müller Tube and this is recorded on the counter.

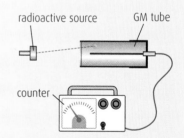

radioactive source   GM tube

counter

A Geiger–Müller tube and counter

Film badges

When working regularly with radioactive sources, employees sometimes wear **film badges**. These badges contain photographic film that darkens when exposed to radiation of all kinds. (Visible light cannot reach the film.) A simplified badge is shown in the diagram.

This badge has three windows.

By analysing where the film has darkened, exposure to different types of radiation can be identified. The amount of darkening of the film also indicates the amount of the radiation received.

A film badge measures exposure to radiation.

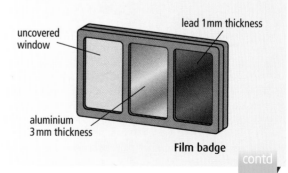

uncovered window

lead 1mm thickness

aluminium 3mm thickness

Film badge

contd

This symbol is used to label radioactive material.

MEASURING BACKGROUND RADIATION

Background radiation can be measured using a timer and a Geiger-Müller tube connected to a counter. Without any radioactive source present, use the timer and counter to measure the number of counts in one minute. Repeat this process several times and take the average of the results to obtain the background count rate.

BACKGROUND RADIATION SOURCES

Everyone is exposed to natural and artificial (man-made) ionising radiation.

Natural radiation is produced from radioactive materials in the land, sea and air. For example, radon gas is released into the atmosphere from soil and rocks. This gas is radioactive and is a major source of naturally occurring radiation. Most (around 80%) of the annual dose of background radiation received by humans is due to natural radiation from cosmic and terrestrial sources. In buildings where monitored levels of radon gas exceed action levels, additional ventilation is installed to remove the gas from the building. Exposure to cosmic rays increase with altitude. The annual flight logs of commercial aircraft pilots and crew who regularly fly at very high altitudes are assessed to predict any excessive exposure to cosmic radiation.

Artificial (or man-made) sources of ionising radiation include radiation from historic nuclear weapons testing in the atmosphere, nuclear power station waste and accidental release, and medical exposure. Medical use of radiation accounts for most of the human exposure to artificial radiation, mainly from X-rays used in diagnosis or treatment of medical conditions. These radiations are described as **background radiation**, because they exist permanently around us. The level of background radiation around the country varies depending on geographic location.

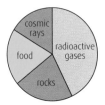

natural sources

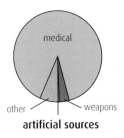

artificial sources

MEASURING EXPOSURE TO RADIATION

Exposure to ionizing radiation can happen under different circumstances:

- as a member of the public – at home, in public, or in a medical situation

- as a worker in the radiation industry.

Exposure to ionising radiation can cause damage to human tissue or organs. Serious acute damage to the body can occur with large exposure to radiation over a short time, but even lower doses of radiation over longer periods of time can cause severe illness such as cancer. The damage depends on the duration of exposure, the type of radiation received and the actual tissues or organs exposed to the radiation.

Since exposure to ionising radiation can be dangerous, it is important to be able to measure this exposure. The Absorbed Dose (D) is used to determine the energy absorbed by exposure to ionising radiation. The Equivalent Dose (H) is used to determine the potential 'harmful effect' of exposure to a particular radiation. This dose depends on the type of radiation, and the type of human tissue or organ exposed to the radiation.

 ONLINE TEST

Take the 'Nuclear Radiation' test at www.brightredbooks. net/N5Physics

 THINGS TO DO AND THINK ABOUT

Determine the level of background radiation in a variety of places near you.

NUCLEAR RADIATION 3

The key concepts to learn in this topic are:

- use of appropriate relationships to solve problems involving absorbed dose, equivalent dose, energy, mass and weighting factor:

$$D = \frac{E}{m}, H = Dw_r$$

- use of an appropriate relationship to solve problems involving equivalent dose rate, equivalent dose and time:

$$\dot{H} = \frac{H}{t}$$

MEASURING THE EFFECT OF RADIATION

Absorbed dose, D

Absorbed dose

$$D = \frac{E}{m}$$

The absorbed dose is how much energy, E, per kilogram from radiation has been received. It depends on the mass, m, of biological material exposed to the radiation, and the absorbed dose is measured in **grays** (Gy).

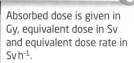
EXAMPLE 1

A 2·5 kg sample of tissue receives 2 mJ of energy. Calculate the absorbed dose.

$$D = \frac{E}{m} = \frac{2 \times 10^{-3}}{2·5}$$

$$= 8 \times 10^{-4} \text{ Gy} = 0·8 \text{ mGy}$$

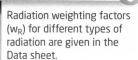

DON'T FORGET

Absorbed dose is given in Gy, equivalent dose in Sv and equivalent dose rate in Sv h⁻¹.

EXAMPLE 2

A worker in a nuclear research plant received an absorbed dose $1·5 \times 10^{-8}$ Gy on a finger when it was exposed to 4·5 µJ of energy from radiation.

Calculate the mass of the worker's finger.

$$D = \frac{E}{m}, \quad 1·5 \times 10^{-6} = \frac{4·5 \times 10^{-8}}{m}, \quad m = 0·03 \text{ kg}$$

Equivalent dose, H

Equivalent dose

$$H = Dw_R$$

The effect of radiation on humans depends on the *absorbed dose*, D, and the type of radiation. Different types of radiation are given a **radiation weighting factor** (w_R) depending on how harmful the effect is on biological material like bone or tissue. **Equivalent dose**, (H), measures this in sieverts, (Sv).

DON'T FORGET

Radiation weighting factors (w_R) for different types of radiation are given in the Data sheet.

The sievert is a very large unit so it is usually expressed in smaller units such as millisieverts (mSv) or microsieverts (µSv).

A table of radiation weighting factors specifies the effect of different types of radiation, for example alpha has a high weighting factor of 20. This is due to the highly ionising effect that alpha radiation has, particularly on contact with skin or if ingested. Note that the table also specifies the ionising effect of other radiations such as fast or slow neutrons and X-rays.

Type of radiation	Radiation weighting factor
alpha	20
beta	1
fast neutrons	10
gamma	1
slow neutrons	3
X-rays	1

contd

EXAMPLE 3

A sample of tissue which received an absorbed dose of 80 μGy was exposed to alpha radiation. Alpha radiation has a radiation weighting factor of 20. Calculate the equivalent dose.

$H = Dw_R = 80 \times 10^{-6} \times 20$ $\qquad\qquad = 1\cdot6 \times 10^{-3}$ Sv $= 1\cdot6$mSv

EXAMPLE 4

A sample of tissue received an absorbed dose of 650 μGy of radiation from a source. The equivalent dose received for this sample of tissue was 1·95 mSv.

Identify the type of radiation used, referring to the table of radiation weighting factors.

$H = Dw_R, 1\cdot95 \times 10^{-3} = 650 \times 10^{-6} \times w_R, w_R = \dfrac{1\cdot95 \times 10^{-3}}{650 \times 10^{-6}} = 3$

so radiation is slow neutrons.

Equivalent dose rate, $\dot{H}$

For safety, it is important to monitor the *rate* at which radiation is absorbed by people who work with radiation. This helps to prevent too much exposure in a short time.

This is known as the **equivalent dose rate** $(\dot{H})$

$\dot{H} = \dfrac{H}{t}$

Equivalent dose rate can be quoted in a variety of units including sieverts, millisieverts or microsieverts per unit time, such as second, minute or hour, for example Sv h^{-1}, mSv h^{-1} and μSv h^{-1}.

EXAMPLE 5

A worker in the nuclear industry received an equivalent dose of 0·02 μSv in 12 hours. Calculate the equivalent dose rate.

$\dot{H} = \dfrac{H}{t} = \dfrac{0\cdot02 \times 10^{-6}}{12} = 1\cdot7 \times 10^{-9}$ Svh^{-1} (Use the same units for time as given in the question.)

EXAMPLE 6

While working with radioactive material, a technician received an equivalent dose of 45 μSv. The equivalent dose rate of exposure to the radiation from the material was 10 μSvh^{-1}.

Determine the duration of the exposure to the radiation.

$\dot{H} = \dfrac{H}{t}, 10 \times 10^{-6} = \dfrac{45 \times 10^{-6}}{t}, t = \dfrac{45 \times 10^{-6}}{10 \times 10^{-6}} = 4\cdot5$ hc

 THINGS TO DO AND THINK ABOUT

1. Investigate the equivalent dose received by aircraft pilots and crew as they routinely fly at altitude above the Earth. Find out how this is monitored.

NUCLEAR RADIATION 4

The key concepts to learn in this topic are:

- comparison of equivalent dose due to a variety of natural and artificial sources
- knowledge of equivalent dose rate and exposure safety limits for the public and for workers in the radiation industries in terms of annual effective equivalent dose:
  - average annual background radiation in UK: 2·2 mSv

- annual effective dose limit for member of the public: 1 mSv
- annual effective dose limit for radiation worker: 20 mSv
- awareness of applications of nuclear radiation: electricity generation, cancer treatment and other industrial and medical uses.

Artificial Source	Annual equivalent dose	
	µSv	mSv
Medical uses (X-rays)	250	0·250
Weapons testing	10	0·010
Nuclear industry (waste)	2	0·002
Other (job, TV, flights)	18	0·018
Total man-made sources	280	0·280

Natural Source	Annual equivalent dose	
	µSv	mSv
Radioactive gases in air and buildings (radon and thoron)	800	0·80
Rocks of the earth	400	0·40
In food and our bodies	370	0·37
Cosmic rays from space	300	0·30
Total natural sources	1870	1·87

COMPARISON OF NATURAL AND ARTIFICIAL SOURCES OF RADIATION

The tables show the annual equivalent dose of radiation received due to artificial and natural sources of radiation.

For the population of the UK, it is estimated that the average annual background radiation in the UK is around 2·2 mSv. This is radiation due to natural sources. Members of the public may be exposed to additional ionising radiations, such as from nuclear fallout from past weapons testing and radioactive waste from nuclear and industrial sites, and some consumer products (for example certain types of smoke detectors).

EFFECTIVE ANNUAL DOSE LIMITS

The **effective annual dose** (in mSv) is a quantity which is calculated to estimate the overall effect of the equivalent dose on different organs of the body.

- For member of the public, the annual effective dose limit is 1 mSv.
- For a radiation worker, the annual effective dose limit is 20 mSv.

These limits are in addition to the normal background radiation for the local environment.

APPLICATIONS OF NUCLEAR RADIATION

Today, the use of nuclear radiation is widespread, for a variety of beneficial reasons. In order to protect people from harmful exposure, any dangers associated with the use of nuclear radiation must be considered and safeguards have to be put into place. The use and applications of nuclear radiation include:

1. electricity generation   2. medical use   3. industrial use.

ELECTRICITY GENERATION

Nuclear reactors harness the heat energy released during the **nuclear fission** of Uranium 235. The heat is used to change water into steam which then drives turbines connected to electricity generators. Nuclear power stations do not produce the gases believed to contribute to global warming. However, the nuclear fuel used is radioactive and, once the fuel is spent, the nuclear waste remains radioactive and requires careful storage to remain safe. Also, nuclear power stations are expensive to decommission (dismantle) once they reach the end of their useful lifetime.

ONLINE

For more details on the key concept of this topic, go to www.brightredbooks.net/N5Physics

ONLINE

For more information about this, check the 'Radiation Therapy for Cancer' link at www.brightredbooks.net/N5Physics

MEDICAL APPLICATIONS OF NUCLEAR RADIATION

Medical research constantly gives rise to new uses of radiation that are adopted throughout the world. Medical applications can be divided into two categories:

1. identification (diagnosis) of health problems
2. health improvement (treatment)

Medical diagnosis

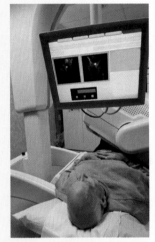

A radioactive **tracer** is a liquid isotope that is injected into a patient. Tracers can be used to check the function and health of various organs of the body.

In the case of a bone scan, a tracer travels through the blood and collects in the bones. The radiation emitted by the tracer is detected by a **gamma camera**.

The images from the camera are analysed by computer and displayed on screens to allow doctors to identify any problems.

Gamma radiation is used for bone scans

Facts about medical tracers

- Tracers must emit gamma radiation – alpha or beta radiation would not penetrate the body to reach the gamma camera outside.
- Any isotope used as a tracer must have a short half-life (see page 100); the activity (see page 94) of the radionuclide must reduce over a short time period so that it does not remain active inside the body for too long (just long enough to permit diagnosis).
- The equivalent dose of the radiation should not be dangerous. (A normal bone scan has the same effect as about 200 X-rays, which doctors consider is not a dangerous amount for this procedure.)
- To produce the tracer, a radioactive substance (called a radionuclide) is attached to a chemical that naturally collects in a particular area of the body. This allows different organs of the body to be targeted for investigation (for example, different chemicals can be labelled to allow doctors to assess the brain, lungs, thyroid gland and kidneys).
- A common radionuclide used in tracers is technetium-99m. This is produced in nuclear reactors.

Medical treatment

External beam radiotherapy is used in the treatment of cancer to destroy cancer cells. Radiotherapy treatment is provided by a machine which emits beams of gamma radiation. The gamma radiation *kills* cancer cells but only *damages* healthy cells, which are then able to recover.

During treatment, the machine containing the gamma source can be rotated around the patient. This directs beams from many angles to target the tumour very precisely. The precision of the instrument means that the surrounding tissue is not exposed to the same large dosage of gamma radiation as the tumour.

Facts about radiotherapy

- The radioactive source commonly used to emit gamma radiation is cobalt-60.
- Cobalt-60 is sometimes produced in nuclear reactors or linear accelerators.
- Other types of radiation used in radiotherapy include high energy X-rays, protons and electrons.

gamma source

Treatment with gamma radiation

INDUSTRIAL APPLICATIONS OF NUCLEAR RADIATION

Radioactive sources are used in many different industrial processes.

Inspecting welds

For example, heavy duty welding to join two pieces of thick steel is common in many industries and it is used in tasks from joining oil pipes to building ships. It is important to be able to check any welded joint for flaws which could cause failure. Air bubbles and other weaknesses are common problems. A gamma source is placed on one side of the welded joint. A detector on the other side picks up differences in the radiation level caused by the presence of air bubbles or other weaknesses.

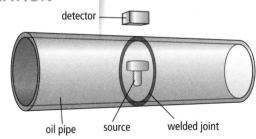

detector

oil pipe　source　welded joint

Using a gamma radiation source to inspect a weld

 THINGS TO DO AND THINK ABOUT

Carry out research into other medical and industrial uses of radiation.

NUCLEAR RADIATION 5

The key concepts to learn in this topic are:
- the definition of *half-life*
- the use of graphical or numerical data to determine the half-life of a radioactive material
- description of an experiment to measure half-life of a radioactive material.

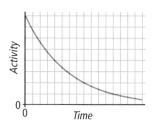

All radioactive sources give graphs of a similar shape

HALF-LIFE

Radioactive decay is a random process – a radioactive substance contains many nuclei of atoms which decay at random. It is impossible to predict *when* the nucleus of an individual atom will decay but, because there are so many nuclei in even a small sample, we can predict the *average number* that will decay in a certain time. Some sources take billions of years to decay while others remain active for only fractions of a second.

Once the atoms in a sample start to decay, there are fewer atoms left to decay. The sample loses *half* its activity after a certain time. This time is known as its half-life. The sample's activity drops by half after each further half-life time.

DON'T FORGET ✚

The time taken for the activity of a substance to halve is called the half-life.

DEFINITION OF HALF-LIFE

The half-life of a radioactive substance is the time for the activity of the substance to decrease by half.

This is the time taken for half the nuclei in the substance to decay.

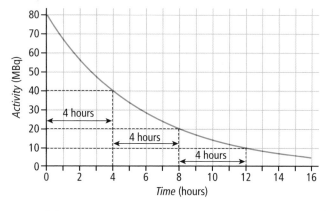

The activity of a radioactive substance decreases with time. If a graph of activity of the radioactive substance against time is plotted, a curve like that shown in the diagram to the left is obtained.

This type of graph can be used to determine the half-life of the radioactive source. The time for the activity to fall from one value to *half* of that value is obtained using values from the graph, as shown in the diagram to the right.

Using the graph, the time for activity to fall from 80–40 MBq is 4 hours. The same time is obtained between 20–10 MBq. The half-life of the source is 4 hours.

EXPERIMENT TO MEASURE THE HALF-LIFE OF A RADIOACTIVE MATERIAL

The background count rate must first be determined:

- Without the radioactive source present, use the timer and counter to measure the number of counts in one minute (CPM). Repeat this process several times and take the average of the results. This is the background count rate. (In this example 8 CPM).

- With the radioactive source present, use the timer and counter to measure the number of counts in one minute. This is the reading at the start.

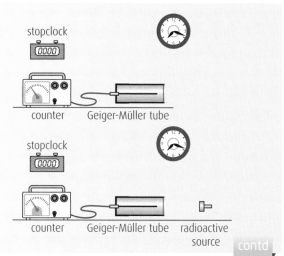

contd

- Measure the count rate per minute at regular intervals. Repeat this process regularly. Record the results in a table.

- Subtract the background count rate from each result to obtain the corrected count rate from the radioactive source alone.

- Draw a graph of Corrected Count Rate against time. Determine the half-life from the graph.

Time (minutes)	Measured count rate (CPM)	Corrected count rate (CPM)	Time (minutes)	Measured count rate (CPM)	Corrected count rate (CPM)
0	248		0	248	240
20	176		20	176	168
40	128		40	128	120
60	92		60	92	84
80	68		80	68	60
100	50		100	50	42

EXAMPLE 1

A Geiger Muller tube connected to a counter is used to measure the counts per minute of a radioactive source every hour for 20 hours (see diagram).

The radioactive source is placed in a lead container to remove the effect of background radiation. Only readings from the source are detected. The results are recorded in a table.

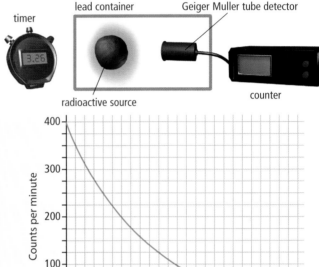

lead container Geiger Muller tube detector

timer

radioactive source

counter

Counts per minute	400	250	160	95	60	33
Time (hours)	0	4	8	12	16	20

A graph of count rate per minute against time is drawn.

Determine the half-life of the source from the graph.

From the graph, the count rate per minute falls from 400 to 200 in 6 hours.

So half-life is 6 hours.

EXAMPLE 2

The activity of a radioisotope used in a hospital was 160 kBq at 6 am on 3rd February. At 6 am on 5th February, its activity was 10 kBq. Calculate the half-life of the radioisotope.

First calculate the number of half-lives needed to get from 160 kBq to 10 Bq.

$160 \rightarrow 80 \rightarrow 40 \rightarrow 20 \rightarrow 10$

This is four halving processes, so four half-lives. From 6 am on 3rd Feb to 6 am on 5th Feb is 48 hours. So there have been four half-lives in 48 hours.

Half-life $= \frac{48}{4} = 12$ hours

SHIELDING OF RADIATION

One of the simplest methods of reducing the danger from ionising radiation is by **shielding**, that is placing an absorbing material in the path of the radiation. Aluminium, lead, concrete of different thicknesses and water have all been used to absorb radiation. Medical radiologists wear lead aprons.

Increasing the distance from the source can also reduce the equivalent dose rate from ionising radiations.

 THINGS TO DO AND THINK ABOUT

Investigate the half-life of the radioactive fuel used in nuclear power stations.

DON'T FORGET

When using a graph to determine half-life, choose values which are easy to read.

ONLINE TEST

Take the 'Nuclear Radiation' test at www.brightredbooks.net/N5Physics

Shielding

NUCLEAR RADIATION 6

The key concepts to learn in this topic are:

- qualitative description of fission, chain reactions, and their role in the generation of energy
- qualitative description of fusion, plasma containment, and their role in the generation of energy.

NUCLEAR FISSION

In **fission**, a nucleus with a large mass number (and so a large number of protons and neutrons in its nucleus) splits into two nuclei of smaller mass numbers, usually with the release of neutrons. Energy is also released.

The amount of energy produced can be huge if a large enough number of atoms are allowed to split in a short time. When atoms of elements like uranium and plutonium split, two entirely different elements are produced, each with smaller atomic masses.

Spontaneous fission

The fission, or decay, of the atoms may be spontaneous. This random nature means that the fission events cannot be reliably predicted.

Spontaneous fission of fermium Fm atoms produces smaller atoms of palladium Pd and xenon Xe, and four neutrons are liberated in the process.

Induced fission

Large atoms can be forced to split, by bombarding them with slow moving neutrons. This process is known as *induced* fission.

In this example of a fission reaction, a uranium U atom is bombarded by a slow neutron, which causes it to split and release energy. Two new smaller atoms of the elements barium Ba and krypton Kr are produced. Three neutrons are also released.

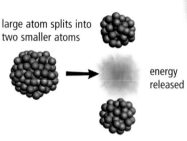

large atom splits into two smaller atoms

energy released

The large atom splits into two atoms of two different elements

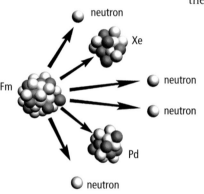

neutron

Xe

Fm

neutron

neutron

Pd

neutron

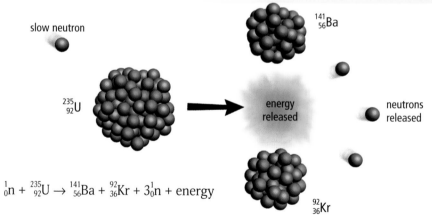

slow neutron

$^{141}_{56}$Ba

$^{235}_{92}$U

energy released

neutrons released

$^{92}_{36}$Kr

$$^{1}_{0}n + {}^{235}_{92}U \rightarrow {}^{141}_{56}Ba + {}^{92}_{36}Kr + 3{}^{1}_{0}n + energy$$

Fission of uranium can be induced

The new elements produced as a result of nuclear fission can vary depending on which large element is used and the initial conditions before the fission.

In a controlled process, the neutrons released in each fission process can be used to bombard further large atoms. These also split, releasing more energy and neutrons. This is called a **chain reaction**.

The energy released by these fission reactions can be very large. It can be removed from the radioactive mass to provide a huge source of energy. Induced fission is used in the reactors of nuclear power stations to provide energy. The energy is used to heat water to produce steam which is used in the generation of electricity.

DON'T FORGET

Energy is released in nuclear fission reactions.

DON'T FORGET

Fission = splitting, fusion = joining.

contd

NUCLEAR FUSION

Fusion is the joining of nuclei.

In fusion, two nuclei combine to form a nucleus of larger mass number. The nuclei that fuse together are usually very small.

A large amount of energy is released, and no radioactive waste is produced in the reaction.

There is a virtually unlimited amount of the isotopes of hydrogen needed for fusion in seawater, and no greenhouse gases are emitted.

$$_1^2H + {}_1^3H \rightarrow {}_2^4He + {}_0^1n + energy$$

Nuclear fusion is the energy source of the sun and the stars. Combining two smaller nuclei to form a nucleus of larger mass number is difficult to achieve because the nuclei consist of positively charged protons which repel each other. Extreme temperatures (1×10^8 K) are required to produce the same fusion reaction that occurs in the sun. At such high temperatures, atoms and molecules lose their electrons and become positively charged ions, to form a **plasma**. When the nuclei are close enough, a strong nuclear force overcomes the repelling electrical force.

The conditions for the **plasma** are it must be:

- heated to a high temperature

- contained inside a container without touching the walls (otherwise the walls would melt)

- confined to a dense plasma to allow high energy output.

The energy released from the fusion will be used to heat water and produce steam for electricity generation using conventional turbines and generators.

Joint research is being carried out by international groups to develop the technology required to produce energy from nuclear fusion.

ONLINE TEST

Take the 'Nuclear Radiation' test at www.brightredbooks.net/N5Physics

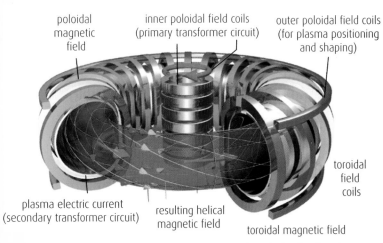

deuterium

tritium

helium

energy

neutron

poloidal magnetic field

inner poloidal field coils (primary transformer circuit)

outer poloidal field coils (for plasma positioning and shaping)

toroidal field coils

plasma electric current (secondary transformer circuit)

resulting helical magnetic field

toroidal magnetic field

An impression of what the containment vessel might be for the plasma – sometimes called a 'magnetic bottle'

THINGS TO DO AND THINK ABOUT

1. Nuclear fission: carry out research into chain reactions. What use is made of this phenomenon in the modern world?

2. Nuclear fusion:
   (a) Carry out research to find out what happens when nuclear fusion runs out in stars.
   (b) Find out what problems there are on Earth which have so far prevented the production of energy by nuclear fusion.

ONLINE

Have a look at the 'Culham Centre for Fusion Energy' link to learn more at www.brightredbooks.net/N5Physics

EXAM PRACTICE

USEFUL TIPS FOR THE EXAM

ONLINE

Head to the BrightRED Digital Zone to find links to print the N5 Data sheet and Relationship sheet.

THE QUESTION PAPER

The question paper for National 5 Physics is worth 135 marks. This is made up of a two hour and 30 minutes question paper with two sections:

- In Section one 25 marks are allocated to an 'objective test' of 25 multiple-choice questions.
- In Section two 110 marks are allocated to the 'written paper' which includes short-answer (restricted) and extended-answer questions.

A Data sheet is provided in the question paper (and can also be found on page 110 of this book). A Relationship list is also provided with this paper (turn to page 109 for a copy). We suggest that you print out a copy of each of these sheets from the BrightRed Digital Zone and keep them beside you when you are doing practice questions. It will help you to become familiar with the various tables and quantities listed.

SECTION 1 (OBJECTIVE TEST)

This has 25 multiple-choice questions worth 1 mark each. These questions are taken from all six areas of the course (approximately six or seven questions per unit). The number of marks for questions from each chapter is approximately in proportion to the content or size of each chapter. Dynamics and Electricity are the longest chapters and usually have more questions than the other chapters. You will be asked to do calculations, recall facts and knowledge from the course, select and analyse data from graphs and tables, and solve problems.

The following are examples of some of the different types of question you may come across.

Standard recall of factual knowledge:

Which of the following contains two vector quantities?

A   Displacement and velocity
B   Weight and speed
C   Distance and speed
D   Force and mass
E   Displacement and speed

This type of question tests your knowledge and memory. Be sure to memorise the knowledge required for the course.

SELECTION OF CORRECT STATEMENTS

A student makes the following statement about different types of waves.

I   Water waves are longitudinal waves
II   Light waves are transverse waves
III Sound waves are longitudinal waves

Which of these statements is/are correct?

A   I only
B   I and II only
C   I and III only
D   II and III only
E   I, II, III

Consider each statement in turn, put a tick or cross beside each statement to make selection of the correct combination easier.

Selection of information to apply to a relationship:

A student sets up the apparatus as shown to measure the average speed of a trolley.

The trolley is released from X and moves down the ramp. The following measurements are recorded.

length of card = 0·06 m
distance from X to Y = 0·40 m
time for card to pass through light gate = 0·04 s
time for trolley to travel from X to Y = 0·25 s
The average speed of the trolley between X and Y is

A   0·06 ms⁻¹
B   0·10 ms⁻¹
C   1·5 ms⁻¹
D   1·6 ms⁻¹
E   10 ms⁻¹.

Use $\bar{v} = \dfrac{d}{t}$,

choose d = 0·4 m,
t = 0·25 s

This requires careful separation of the correct values from non-relevant information to calculate the answer.

contd

SELECTION FROM A TABLE OF ANSWERS

A uniform electric field exists between metal plates X and Y. The diagram shows the path of a charged particle as it passes through the field.

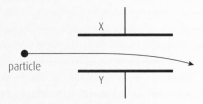

Which row in the table identifies the charge on the particle, the charge on plate X and the charge on plate Y.

	Charge on particle	Charge on plate X	Charge on plate Y
A	positive	negative	positive
B	no charge	positive	negative
C	negative	negative	positive
D	no charge	negative	positive
E	negative	positive	negative

Take time to consider each row – mark wrong answers with a cross to help identify the correct row.

ANALYSIS OF INFORMATION IN A DIAGRAM

The following diagram gives information about a wave.

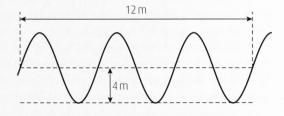

Which row shows the wavelength and amplitude of the wave?

	Wavelength (m)	Amplitude (m)
A	2	4
B	2	8
C	3	4
D	3	8
E	4	8

Carefully consider the information given in the diagram to find the values required.

SELECTION AND APPLICATION OF CORRECT RELATIONSHIP

Four resistors are connected as shown.

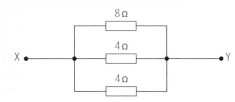

A	0.06Ω
B	0.6Ω
C	1.6Ω
D	10Ω
E	16Ω

Select the relationship required from the relationship sheet, write down you working on the rough working paper.

SECTION 2

There are a total number of 110 marks in section 2. This consists of about 12–14 questions of different types. These questions are taken from all six chapters of the course. The number of marks for questions from each chapter is approximately in proportion to the content or size of each chapter.

The following are examples of the different types of questions which could be asked in this exam. Specimen answers are given.

Questions based upon course content

This type of question usually consists of several parts and is based on applying your knowledge and skills, from one or more key areas of the course units, to a question.

Example 1

(a) A student sets up a circuit to operate three identical 6V, 18W lamps from a 30V supply.

When the switch is closed, the lamps operate at their correct power rating.

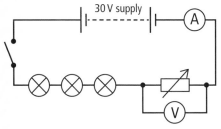

Calculate:

(i) The reading on the ammeter;

$$P = IV \qquad 18 = I \times 6 \qquad I = \frac{18}{6} = 3A$$

(ii) The reading on the voltmeter;

$$30 = 6 + 6 + 6 + V_R \qquad V_R = 30 - 18 = 12V$$

(iii) The resistance of the variable resistor. 3

$$V = IR \qquad 12 = 3 \times R \qquad R = \frac{12}{3} = 4Ω$$

(b) The student sets up a second circuit using a 6V supply and two of the same lamps. Each lamp has a resistance of 2Ω. The resistance of the variable resistor is set to 6Ω.

contd

(i) Calculate the total resistance of this circuit. **3**

$$\frac{1}{R_T} = \frac{1}{R_1} + \frac{1}{R_2} + \frac{1}{R_3} \qquad \frac{1}{R_T} = \frac{1}{2} + \frac{1}{2} + \frac{1}{6}$$

$$\frac{1}{R_T} = 0.5 + 0.5 + 0.167 = 1.167$$

$$R_T = \frac{1}{1.167} = 0.9\,\Omega$$

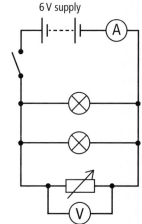

(ii) One lamp is now removed from the circuit.
(A) What happens to the reading on the ammeter? **1**

The reading decreases.

(B) Justify your answer. **1**

The resistance increases so the current decreases.

Questions based on practical or experimental work

This type of question requires you to apply your knowledge and skills to a practical or experimental situation. Usually, there is a brief explanation of an experiment with a diagram, which can be followed by a table of results or a graph (or both). You may be required to take information from the table or graph and to calculate answers. Sometimes, the final part of the question asks for a suggested improvement to the experiment.

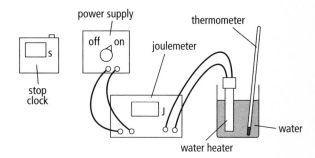

Example 2
(a) A student carried out an experiment to determine the specific heat capacity of water.
The energy supplied to the water was measured using a joulemeter.
The following information was recorded:
• Initial temperature of the water = 25°C
• Final temperature of the water = 35°C
• Initial reading on the joulemeter = 16 kJ
• Final reading on the joulemeter = 148 kJ
• Mass of water = 3.0 kg
• Time = 11 minutes

(i) Calculate the change in temperature of the water. **1**

$$\Delta T = (35 - 25) = 10°C$$

(ii) Calculate the energy supplied by the water heater. **1**

$$E_{supplied} = (148\,000 - 16\,000) = 132\,000\,J$$

(iii) Calculate the value for the specific heat capacity of water obtained from this experiment. **3**

$$E_h = cm\Delta T \qquad 132\,000 = c \times 3.0 \times 10$$
$$c = 4400\,J\,kg^{-1}\,°C^{-1}$$

(b) (i) The accepted value for the specific heat capacity of water is quoted in the table in the Data sheet. Explain the difference between the accepted value and the value obtained in the experiment. **2**

Since $c = \dfrac{E_h}{m\Delta T}$ The measured value of E_h is too large
OR the water is not evenly heated
OR temperature difference ΔT is too small
OR heat is lost to the surroundings.

(ii) How could the experiment be improved to reduce this difference? **1**

Insulate the beaker OR put the lid on the beaker OR stir the water.

(c) Calculate the power rating of the water heater. **3**

$$E = Pt \qquad 132\,000 = P \times 11 \times 60 \qquad P = 200\,W$$

Questions based on content not specified within the course

This type of question usually consists of a description of an aspect of physics that is not covered in the National 5 course, including presentation of a new relationship with each variable defined. Part of the question will usually ask for a value to be calculated using the new relationship. Other parts of the question may ask about the normal coursework.

Example 3
When a glass tube with a very narrow hollow centre is placed in a trough of water, a property called **surface tension** causes some water from the trough to rise up into the tube. This can be used to determine a value for the surface tension.

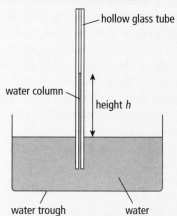

contd

Surface tension, symbol γ, is measured in $N\,m^{-1}$.

γ for water using this method is given by:

$$\gamma = \frac{rh\rho g}{2}$$

Where: r is the internal radius of the glass tube in metres
h is the height of the water column inside the glass tube in metres
ρ is the density of the water in $kg\,m^{-3}$
g is the acceleration due to gravity in $m\,s^{-2}$

The results when one particular glass tube was used were:
• the internal radius was 0·15 mm
• the height of water in the column was 0·1 m
• the density of water was 1000 $kg\,m^{-3}$
• g = 9·8 $m\,s^{-2}$

(a) Calculate the value for the surface tension, in $N\,m^{-1}$, obtained from these results. **2**

$$\gamma = \frac{rh\rho g}{2} = \frac{0\cdot15 \times 10^{-3} \times 0\cdot1 \times 1000 \times 9\cdot8}{2} = 0\cdot0735\ N\,m^{-1}$$

(b) The total mass of the water trough and glass tube is 3·5 kg. The area of the trough in contact with the laboratory bench is 0·07 m^2. Show that the pressure exerted by the trough on the bench is 490 Pa. **4**

$$W = mg = 3\cdot5 \times 9\cdot8 = 34\cdot3\ N$$

$$p = \frac{F}{A} = \frac{34\cdot3}{0\cdot07} = 490\ Pa$$

Questions based on content within the course, assessing skills

This type of question usually consists of a description of an experiment. It may also have a table of results or graph. You may be asked to select information from the data given, and part of the question may ask for a suggestion or a prediction.

Example 4
A student releases a ball from rest near the top of a track. The ball moves down the track and passes through a light gate near the bottom of the track.

The student records the following information:
• Distance travelled by the ball down the track = 1·4 m
• Diameter of ball = 40 mm
• Time for the ball to pass through the light gate = 0·065 s

(a) Calculate the speed of the ball as it passes the light gate. **3**

$$\text{speed} = \frac{d}{t} = \frac{\text{diameter of ball}}{t} \qquad \text{speed} = \frac{40 \times 10^{-3}}{0\cdot065} = 0\cdot62\ m\,s^{-1}$$

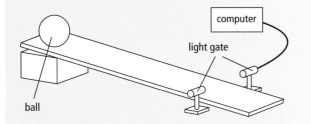

(b) The mass of the ball is 0.16 kg. Calculate its kinetic energy as it passes through the light gate. **3**

$$E_k = \tfrac{1}{2}\,mv^2 = \tfrac{1}{2} \times 0\cdot16 \times (0\cdot62)^2 = 0\cdot03\,J$$

(c) Suggest a possible value for the average speed of the ball over the 1·4 m distance travelled by the ball down the track. **1**

$$0\cdot3\ m\,s^{-1}$$

An open ended question (3 marks)

This question usually describes an application or event and is followed by a statement. You are asked to think about this application and use your knowledge of physics to comment on the statement. The marks awarded depend on the depth of your answer. You need to think about which areas of physics are important and set out your answer in a sensible form that makes sense to the reader.

Example 5
During a game of squash, the players strike the ball with a racquet. The ball then rebounds against a wall. Rallies can consist of several returns. The ball is made of a rubber compound, filled with air.

After a match, one player observes that the squash ball is harder to compress than before the match. Use your knowledge of physics to comment on this observation.

When the ball collides with the wall or racquet some of its kinetic energy is transformed into heat energy. This heat energy heats up the air inside the ball. The volume does not change.

As the temperature of the air inside the ball increases, the pressure of the air inside the ball increases. So greater pressure is needed to compress the ball.

Questions assessing scientific literacy

These questions usually present a passage of scientific information. You may already know about some of the information, but some of it may be new to you. The questions may require you to interpret the passage, or to make a prediction based on what you have read.

Example 6
Read the passage below and answer the questions that follow.

Galaxies are composed of dust, gas, and millions of stars and planets. There are too many galaxies to be counted. The universe which can be detected today has more than 90 billion galaxies.

contd

There are three types of galaxy:

- Elliptical: these galaxies are 'egg' shaped. They are generally round but stretched more in one direction than the other. They may be nearly circular, or so 'squashed' that they take on a tubular appearance. Elliptical galaxies are the largest known galaxies and contain many stars. The number can be more than one million, million stars! However, elliptical galaxies contain very few dust or gas clouds, where new stars are formed. So, there are few new stars in elliptical galaxies. Messier 87 is a huge elliptical galaxy, measuring some 120 000 light years in diameter. Stars in elliptical galaxies loosely orbit the centre of the galaxy.

- Spiral: these consist of a flat saucer-shaped disc with a spherical centre which has arms 'spiralling' out from it like a 'Catherine wheel' firework. These galaxies contain stars, dust, gas and planets.
  In spiral galaxies, all of the stars and interstellar material orbit the centre of the galaxy.
  The Milky Way is an example of a small spiral galaxy. New stars are continually being formed in the dust clouds of spiral galaxies.

- Irregular: these have no recognisable shape or features. Some irregular galaxies have a peculiar, misshapen appearance, and contain mostly new stars. Some irregular galaxies are the result of two or more galaxies colliding, and contain a large amount of dust and gas.

(a) Which type of galaxy is the largest?   1

    Elliptical galaxies.

(b) Why do elliptical galaxies contain few new stars?   1

    New stars are formed in dust and gas clouds – elliptical galaxies contain little dust and few gas clouds.

(c) Show that the diameter of Messier 87 is $1\cdot14 \times 10^{21}$ m.   3

    $d = vt$
    $d = 3 \times 10^8 \times (365\cdot25 \times 24 \times 60 \times 60) \times 120\,000$
    $d = 1\cdot14 \times 10^{21}$ m

Questions based on an application of the course content

This type of question usually consists of several parts and asks you to apply knowledge and skills. There is usually an introduction (with a diagram) which describes the application.

The application usually contains a description of a machine or a system which is commonly used. It may even refer to the latest technology or to a new invention. You may be asked to calculate or deduce answers based on your knowledge of the coursework.

Example 7

A halogen heater contains two heater tubes which can be switched on separately. The heater also has an 'uplighter' lamp that can be switched on to illuminate the ceiling.

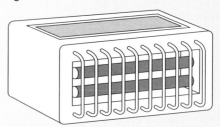

The heater has three different heat settings: LOW, MEDIUM and HIGH. These settings can be produced by switching on the heating tubes.

The circuit diagram for the heater is shown.

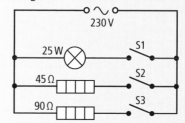

(a) When switch S1 is closed, the lamp operates at its rating of 25 W.
Calculate the current in the lamp.   3

$$I = \frac{P}{V} = \frac{25}{230} = 0\cdot11\,\text{A}$$

(b) Switch S1 is opened and switches S2 and S3 are closed.

(i) Calculate the combined resistance of both heater tubes.   3

$$\frac{1}{R_T} = \frac{1}{R_1} + \frac{1}{R_2} \qquad \frac{1}{R_T} = \frac{1}{45} + \frac{1}{90} \qquad R_T = 30\,\Omega$$

(ii) Calculate the total power developed in the heating elements when S2 and S3 are closed.   3

$$P = \frac{V^2}{R} = \frac{230^2}{30} = 1763\cdot3\,\text{W}$$

(iii) State and explain which switch or switches would have to be closed to produce the LOW heat setting.   3

S3 (only).
Greatest value of resistance gives lowest current.
Since $P = I^2 R$ this gives lowest power output

RELATIONSHIPS REQUIRED FOR NATIONAL 5 PHYSICS

$d = vt$

$d = \bar{v}t$

$s = vt$

$s = \bar{v}t$

$a = \dfrac{v - u}{t}$

$F = ma$

$W = mg$

$E_w = Fd$

$E_p = mgh$

$E_k = \dfrac{1}{2}mv^2$

$Q = It$

$V = IR$

$V_2 = \left(\dfrac{R_2}{R_1 + R_2}\right)V_s$

$\dfrac{V_1}{V_2} = \dfrac{R_1}{R_2}$

$R_T = R_1 + R_2 + \ldots$

$\dfrac{1}{R_T} = \dfrac{1}{R_1} + \dfrac{1}{R_2} + \ldots$

$P = \dfrac{E}{t}$

$P = IV$

$P = I^2R$

$P = \dfrac{V^2}{R}$

$E_h = cm\Delta T$

$E_h = ml$

$p = \dfrac{F}{A}$

$p_1 V_1 = p_2 V_2$

$\dfrac{p_1}{T_1} = \dfrac{p_2}{T_2}$

$\dfrac{V_1}{T_1} = \dfrac{V_2}{T_2}$

$\dfrac{pV}{T} = \text{constant}$

$f = \dfrac{N}{t}$

$v = f\lambda$

$T = \dfrac{1}{f}$

$A = \dfrac{N}{t}$

$D = \dfrac{E}{m}$

$H = Dw_r$

$\dot{H} = \dfrac{H}{t}$

DATA SHEET

Speed of light in materials

Material	Speed in ms^{-1}
Air	3.0×10^8
Carbon dioxide	3.0×10^8
Diamond	1.2×10^8
Glass	2.0×10^8
Glycerol	2.1×10^8
Water	2.3×10^8

Gravitational field strengths

	Gravitational field strength on the surface in Nkg^{-1}
Earth	9.8
Jupiter	23
Mars	3.7
Mercury	3.7
Moon	1.6
Neptune	11
Saturn	9.0
Sun	270
Uranus	8.7
Venus	8.9

Specific latent heat of fusion of materials

Material	Specific latent heat of fusion in Jkg^{-1}
Alcohol	0.99×10^5
Aluminium	3.95×10^5
Carbon dioxide	1.80×10^5
Copper	2.05×10^5
Iron	2.67×10^5
Lead	0.25×10^5
Water	3.34×10^5

Specific latent heat of vaporization of materials

Material	Specific latent heat of vaporisation in Jkg^{-1}
Alcohol	11.2×10^5
Carbon dioxide	3.77×10^5
Glycerol	8.30×10^5
Turpentine	2.90×10^5
Water	22.6×10^5

Speed of sound in materials

Material	Speed in ms^{-1}
Aluminium	5200
Air	340
Bone	4100
Carbon dioxide	270
Glycerol	1900
Muscle	1600
Steel	5200
Tissue	1500
Water	1500

Specific heat capacity of materials

Material	Specific heat capacity in Jkg^{-1} °C
Alcohol	2350
Aluminium	902
Copper	386
Glass	500
Ice	2100
Iron	480
Lead	128
Oil	2130
Water	4180

Melting and boiling points of materials

Material	Melting point in °C	Boiling point in °C
Alcohol	−98	65
Aluminium	660	2470
Copper	1077	2567
Glycerol	18	290
Lead	328	1737
Iron	1537	2737

Radiation weighting factors

Type of radiation	Radiation weighting factor
Alpha	20
Beta	1
Fast neutrons	10
Gamma	1
Slow neutrons	3
X-rays	1

GLOSSARY

absorbed dose (*D*)   The amount of energy per kilogram (of human tissue) received from exposure to nuclear radiation.

absolute zero   The lowest temperature on the kelvin temperature scale, 0 kelvin (−273 °C).

absorption   When materials absorb the energy of nuclear radiation.

Acceleration, *a* (m s⁻²)   The rate of change of velocity (positive or negative) in unit time, calculated using $a = (v - u)/t$.

activity (becquerels, Bq)   The number of atoms in a radioactive substance that disintegrate per unit time.

air resistance   A frictional force acting on objects which move through air (sometimes called 'drag').

alpha particle, alpha radiation   Radiation caused by a relatively large, slow moving radioactive particle with a positive charge (a helium nucleus).

alternating current (a.c.)   The periodic movement of electric charge between the supply terminals.

ammeter   An instrument for measuring electric current.

amperes (A)   Unit of electric current, also amps.

amplitude   The peak value of an alternating quantity, e.g. a wave. The height of a wave measured from the mid position to the crest (or trough) of the wave.

angle of incidence   The angle that a ray of light makes with the normal before it passes into a medium, e.g. glass.

angle of refraction   The angle that a ray of light makes with the normal when it passes into a medium, e.g. glass.

atom   A basic unit of matter, composed of protons, neutrons, and electrons.

average speed, $\bar{v}$ (m s⁻¹)   The speed of an object over a relatively long period of time or distance.

average velocity, $\bar{v}$ (m s⁻¹)   The speed and direction of an object over a relatively long period of time or distance.

background radiation   Radiation in the atmosphere due to various natural and man-made radioactive sources.

battery   Two or more electric cells which transfer electrical energy to charges.

becquerels (Bq)   The unit of activity of a radioactive substance where one becquerel is one decay (of an atom) per second.

beta particle, beta radiation   Radiation caused by a relatively small, fast moving radioactive particle with a negative charge (electron).

big bang model/theory   Current theory of the formation of the universe.

black hole   A space where the gravitational force is so strong that even light cannot escape, sometimes caused when stars collapse.

capacitor   An electronic component which stores charge.

chain reaction   A neutron hits a uranium nucleus causing it to split (fission). This produces heat energy and more neutrons, which in turn cause more fission events to take place.

chemical energy   Energy produced from chemical compounds, including coal, oil and gas.

concave   The shape of lens that causes light rays to diverge.

condensation   When a gas changes state into a liquid.

conductor   A material which allows charge to move through it easily.

conservation of energy   The total energy before and after a transformation from one form into another is unchanged.

convex   The shape of lens that causes light rays to converge to a focus.

cosmic rays   High energy particles which arrive at Earth from space.

crest   The topmost part of a wave shape.

current (*I*)   The flow of charge per unit time measured in amperes.

deceleration (m s⁻²)   The rate of change of velocity, i.e. change in velocity divided by the time taken when an object is slowing down.

diffraction   A property of waves whereby they 'bend' around a gap or obstacle.

diode   An electrical component which allows current through it in one direction only.

direct current (d.c.)   The movement of charge which is always in one direction between the terminals of the supply.

displacement   The direct distance of a finishing point from a starting point, including the direction.

electrical energy   Type of energy associated with electric charge.

electromagnetic (EM) spectrum   Waves which range from long wavelength radio waves to gamma with short wavelengths. They all travel at the speed of light (3×10^8 m s⁻¹).

electromagnetic waves   The waves of the electromagnetic spectrum.

energy (joules, J)   The ability of an object or system to do work.

energy transformation   Describes how energy from one source is changed into another energy source, e.g. $E_p \rightarrow E_k$.

equivalent dose (*H*)   A measure of the effect that exposure to radiation has on humans, measured in Sieverts (Sv).

equivalent dose rate ($\dot{H}$)   A measure of the rate of absorption of nuclear radiation by humans.

filament lamp   A type of lamp that uses a metal resistance wire to transform electrical energy to light (and heat).

film badge   A monitoring device worn by people who work with radioactive materials to record the type and quantity of any exposure to radiation.

fission   The process of splitting the nucleus of an atom into smaller nuclei and releasing energy.

force (N)   A pull or a push, measured in newtons. One newton is the force required to give a mass of one kilogram the acceleration of 1 m s⁻².

freefall   Describes the motion of an object falling freely due to the force of gravity.

frequency (hertz, Hz)   The number of events per second or waves which pass a point in one second.

friction   A force that acts in the opposite direction to objects as they move or try to move.

fusion (1)   This is the process of joining two smaller atoms (or nuclei) to produce a larger atom with the release of energy.

(2)   The change of state of a substance from solid to liquid.

gamma camera   A device containing a detector of gamma radiation.

gamma radiation, gamma rays   A group of waves emitted by some radioactive materials, part of the EM spectrum.

Geiger–Muller tube   A radiation detector, usually connected to a counter device, which can detect alpha, beta and gamma radiation.

geostationary orbit   An orbit above the equator in which a satellite takes 24 hours to orbit the Earth.

gravitational field strength (*g*)   The weight per unit mass (or the force of gravity acting on each kilogram) used to calculate the weight of an object.

gray (Gy)   The unit for absorbed dose.

half-life   The time for the activity of a radioactive substance to reduce (or decay) to half its original value.

infrared (IR) radiation   Also known as heat rays or waves, a group of waves which are part of the electromagnetic spectrum.

interference   A property of waves whereby waves from different sources can combine (not in this course).

ion   An atom which has lost or gained one or more electrons, becoming a charged particle.

ionisation   When nuclear radiation changes atoms into ions.

isotope   Different forms of the same element, with different numbers of neutrons in the nucleus, but with the same atomic number.

kinetic energy (E_k)   Energy an object possesses when moving.

kinetic theory of gases   An explanation of the behaviour of particles in a gas in terms of the volume, pressure and temperature of the gas.

latent heat   The energy required to change the state of a substance.

light year   The distance travelled by light in one year.

longitudinal wave   Waves in which the particles transferring energy vibrate in the direction of travel, e.g. sound waves.

medium   The material which waves travel through, e.g. water, glass.

microwaves   A group of waves which are part of the electromagnetic spectrum with wavelengths that are shorter than those of radio waves.

Newton's laws of motion   First, second and third laws identify the effect of forces on objects.

normal   A line drawn at right angles to an edge (usually of glass), used when drawing ray diagrams to represent the path of light waves.

nuclear fusion   Two nuclei of small mass combine to form a larger nuclei with release of energy.

oscilloscope   An electrical device with a screen that is used to display electrical signals which can demonstrate wave motion.

period (T)   The time taken for one wave to be produced.

potential difference (p.d.)   A measure of how much energy is transferred to charges in a circuit.

potential energy (E_p)   The stored energy of an object; e.g. an object that has been raised above ground level gains (gravitational) potential energy.

prism   Triangular three-dimensional glass block which is used to refract light rays, usually to analyse the frequencies of light present in the rays.

projectile   An object which has been launched (or projected) into the air.

radiation weighting factor (w_R)   A measure of the harmful effect on human tissue of nuclear radiation.

radio signals or waves   Electromagnetic waves which are part of the electromagnetic spectrum with high frequency and short wavelength.

radionuclide   An isotope of a radioactive element.

reflection   A property of waves whereby they return when reaching a reflective barrier.

refraction   A property of waves whereby the waves change speed (and sometimes direction) when passing from one medium into another, e.g. light waves passing from air into glass.

resultant   The result of adding two (or more) vectors.

resistance   A measure of the opposition to the movement of charge in a circuit.

satellite   An object orbiting a star or planet.

scalar   A physical quantity which has size only.

shielding   The process of, or material required, to absorb radiation to prevent exposure of humans to its effects.

sievert (Sv)   The unit for equivalent dose, H.

speed (v)   The distance travelled by an object per unit of time ($m s^{-1}$).

specific heat capacity (c)   The amount of heat energy required to raise the temperature of 1 kg of a substance by 1 °C.

streamlining   The effect of reducing the air resistance (or 'drag') of a moving object, making the object's shape more 'aerodynamic'.

supernova   The result of an explosion of a massive star in space (plural supernovae).

television wave   Electromagnetic waves which are part of the electromagnetic spectrum with high frequency and short wavelength.

terminal velocity   The velocity reached by a moving object when the forces acting on it are balanced.

thrust   Used to describe the force produced by an engine (e.g. rocket engine).

tracer   A radioactive material injected into a human to help with the diagnosis of particular health problems.

transmitter   A device which sends out or produces waves.

transverse wave   Waves where the particles transferring energy vibrate at right angles to direction of travel, e.g. water waves.

trough   The lowest part of a wave shape.

ultraviolet (UV) radiation   A group of waves which are part of the electromagnetic spectrum.

vaporisation   The change of state of a substance from liquid to gas.

vector   A physical quantity which has size and direction.

velocity (v)   The displacement of an object per unit of time, including the direction of the object ($m s^{-1}$).

vibration   A backward and forward motion, which usually produces waves.

visible spectrum   Name given to the range of wavelengths of visible light waves giving a range of colours.

wave equation   An equation linking wavespeed, v, frequency, f, and wavelength, λ: $v = f\lambda$.

wavelength (λ)   The length of one wave, the distance from one position on a stream of waves to the next identical position where a new wave starts (m).

wavespeed (v)   The speed at which waves travel through a medium ($m s^{-1}$).

weight (W)   The gravitational force acting on an object (N).

work done (E_w)   The amount of energy required when a force, F, is applied to move an object a distance, d. ($E_w = Fd$) (J).

X-rays   Fast-moving electrons referred to as waves which are part of the EM spectrum.